# KITCHI

## COOKBOOK
## FOR BEGINNERS

**The Complete Stand Mixer
Cookbook with Easy and
Delicious Recipes to Try**

**Barbara Colbert**

# CONTENTS

# Introduction

**F**or home bakers and professionals alike, a sturdy stand mixer is key. Stand Mixer is every cook's dream machine. It provides an extra pair of hands in the kitchen for busy cooks who need as much help as they can get to create delicious made-from-scratch meals. As much as I like to make my meals and that of my family, I make sure not to spend unneccesary hours doing that. Stand mixers have been helping to reduce time spent for food preparation in professional and domestic kitchens for almost a century.

Stand mixers are designed to mix, whisk and knead perfect breads and cakes while its sturdy attachments do everything from cutting, slicing and shredding to making fresh pasta, juicing fruits or grinding meats.

For the purpose of this cookbook, we will focus on the Kitchen Aid Stand Mixer as this kitchen appliance has been rated as the best stand mixer.

# 1. BEFORE GETTING STARTED

## Stand mixer vs Hand mixer

Besides the fact that you have to hold one while you're using it, the biggest differences between a stand mixer vs. handheld mixer are motor size and price. While a handheld electric mixer is a useful appliance for smaller jobs like mousse or pudding, it requires you to be present, holding it while it runs. A stand mixer allows you some freedom of movement in the kitchen.

A hand mixer is more compact and fits better for small kitchens and tight budgets alike. A stand mixer is definitely a larger investment for the home kitchen. It takes up significantly more room, is heavier, but has the power to easily tackle heavy-duty batter, bread doughs, and baking projects. Typically it'll provide quicker results with less mess.

# What are the best mixer attachments?

**Paddle:** Otherwise known as a flat beater. This is a good multi-purpose tool and should be used for heavier mixtures such as cake batter, frostings, cookies, mashed potatoes, and meatloaf.

**Wire whip:** Like a handheld whisk, the wire whip has stainless steel wires attached to a hub. The whisking motion of the whip incorporates air into eggs, egg whites, whipping cream, candies, angel food cake, and mayonnaise. Not to be used with dough and heavier mixtures.

**Dough hook:** The dough hook can resemble a big "C" or a spiral and it is used for kneading yeast doughs like breads, coffee cakes, pizza dough, cinnamon rolls, and even pasta.

**Scraper:** A beater with a flexible rubber or silicone fitting on one or both edges that scrape down the sides of the mixing bowl as it turns. The silicone edge on the blade can do what your rubber spatula does, so you don't have to stop the mixer mid-way through and do it yourself. Perfect for sticky foods like cream cheese and nut butter.

## Speed settings

Most models will feature 3 to 12-speed settings, but all good stand mixers should have a "slow start" setting, for adding ingredients to the mixture without a big mess. Most bakers can make do with 3-speed settings, but the more options there are, the more accurate your work will be.

Slow speed for starting the machine or combining dry ingredients.

**Speed 2 (low)** for slow mixing heavy batter or cutting in butter.

**Speed 4 (medium-low)** for mixing cookie doughs or beating egg whites.

**Speed 6 (medium)** for creaming butter and beating frostings.

**Speed 8 (medium-high)** for fast beating or whipping to make meringues and whipped cream.

**Speed 10 (high)** for fast whipping a small amount of egg whites or cream.

## Maintenance

For the best results, always make sure that the bottom of the attachment is touching the mixing bowl. This may require some adjustment, depending on your machine. Also periodically check your mixer for loose screws, tightening as necessary.

Vibrations in the machine can loosen them over time. Inspect the bottom of the mixing bowls for undue wear and tear, that would prevent them from locking into the base. They may need replacement when they fit loosely or rattle when attached to the base.

## Cleaning

Most stand mixers come equipped with a stainless steel mixing bowl, but some brands are revisiting glass, copper, or porcelain in their latest versions. All are fine choices, but some may require a little more care than stainless steel. Most mixer parts, including the bowl, can be washed safely in a dishwasher, or by hand.

After every use, go over the stand mixer itself with a damp cloth to clean any spatter on the undersides of the machine. Make sure your mixer, mixing bowl, and attachments are all free of dust and dirt before starting a new project.

# 2.

# BASIC RECIPES

# TOMATO SAUCE

No cook should be without a recipe for a good tomato sauce. It is the basis for so many dishes: pasta sauces, pizza toppings, stews and even soups.

**Serves 4**

**Prep: 5 minutes**

**Cook: 50 minutes**

**2 onions**

**1 celery stalk**

**1 carrot**

**2 garlic cloves**

**100 ml olive oil**

**1 kg ripe plum tomatoes,**

**or 800 g tinned plum tomatoes**

**1 tsp dried oregano**

**¼ tsp chilli flakes (optional)**

**1 bay leaf**

**1 sprig of rosemary**

**2 tbsp tomato purée**

**a pinch of sugar (optional)**

**salt and freshly ground black pepper**

Chop the vegetables and garlic with the medium shredding drum on the rotor vegetable slicer/ shredder on speed 4. Heat the olive oil in a large saucepan and sauté the vegetables and garlic for 5 minutes until softened.

Purée the tomatoes into the mixer bowl with the coarse grinding plate on the food grinder on speed 4. Add to the pan with the herbs and tomato purée. Add a pinch of sugar, if necessary. Season and simmer for 45 minutes until the sauce has thickened and intensified in flavour.

# PASTA DOUGH

Homemade pasta dough is quick and easy to make. Be sure to use Italian '00' flour, as ordinary flour will result in a grey and unappealing dough. This recipe is also the basis for so many other recipes in this book.

**Makes 450 g**

**Prep: 10 minutes**

**Rest: 1 hour**

**300 g '00' flour**

**3 eggs**

**1 tbsp olive oil**

Place the flour, eggs and olive oil in the mixer bowl. Mix slowly with the flat beater on speed 1 until you obtain a homogenous mixture. Change to the dough hook and increase to speed 2, kneading for 5 to 8 minutes until the dough is smooth and elastic. Wrap in clingfilm and chill for 1 hour.

Attach the pasta roller and set the adjustment knob to 1. Cut the dough into 4 portions, roll out with a rolling pin and pass one portion slowly through the roller at speed 2. Fold into three and pass through the roller again, then repeat this process four or five times.

When the dough is no longer sticky, set the adjustment knob to 2 and pass the pasta through. Do not fold the dough this time. Continue rolling the dough through, two or three times on each setting, until you obtain the desired thickness. Repeat with the remaining portions of dough, then cut the dough into the desired shapes.

# MAYONNAISE

Making your own mayonnaise is very rewarding and easy to do. You can vary this basic recipe by experimenting with different flavours. A few suggestions can be found below.

**Makes 250 ml**

**Prep: 10 minutes**

**1 large egg**

**1 egg yolk**

**1 tsp lemon juice, plus extra if needed**

**1 tsp Dijon mustard**

**100 ml olive oil**

**100 ml groundnut oil**

**salt and freshly ground black pepper**

Warm the mixer bowl and wire whisk under hot running water, then dry thoroughly. Make sure all the ingredients are at room temperature, otherwise the mayonnaise will split. Place the egg, egg yolk, lemon

17

juice and mustard in the mixer bowl. Whisk with the wire whisk on speed 8 until well-combined and frothy.

Mix the oils in a measuring jug, then add the oil to the bowl while the motor is running: drop by drop at first, then in a thin stream as the mayonnaise begins to thicken. When all the oil has been absorbed, season the mayonnaise and sharpen with extra lemon juice if necessary. This mayonnaise will not thicken as much as a mayonnaise prepared with only egg yolks.

## Variations

### Wasabi mayo

Stir 1 to 2 tablespoons of wasabi paste into the mayonnaise. Serve with crab cakes.

### Sesame mayo

Replace the olive oil with 75 ml toasted sesame seed oil and 25 ml groundnut oil. Stir in 2 tablespoons of toasted sesame seeds. Serve with white fish and cold poached chicken.

### Anchovy and lemon mayo

Pound half a tin of anchovy fillets to a paste with a mortar and pestle and stir into the mayonaise. Add the grated zest of 1 lemon. Serve with grilled pork or lamb chops, or roast salmon.

# PESTO

Pesto is traditionally flavoured with basil but other herbs work just as well, for example mint, coriander, parsley, sage and even rocket.

**Makes 200 ml**

**Prep: 10 minutes**

**30 g Parmesan cheese**

**30 g Pecorino Romano**

**60 g fresh basil**

**40 g toasted pine nuts**

**2 garlic cloves**

**120 ml extra virgin olive oil**

**salt and freshly ground black pepper**

Grate the cheeses with the fine shredding drum on the rotor vegetable slicer/shredder on speed 4.

Place the basil, pine nuts and garlic in the blender, and blend to a purée on stir speed; make sure that the pesto retains some texture. With the motor running, gradually add the olive oil until all the oil has been absorbed. Scrape the mixture into a bowl and fold in the cheese. Season to taste but be careful with the salt, as the cheese is quite salty. Use immediately, or cover with a thin layer of olive oil and store in the refrigerator.

## Variations

### Mint and pistachio pesto

Replace the Pecorino with Parmesan cheese, the basil with mint and the pine nuts with shelled pistachio nuts. Proceed as described above.

This pesto is delicious with grilled aubergines or roast lamb.

### Rocket and goat's cheese pesto

Replace the cheeses with 120 g fresh goat's cheese and the basil with rocket. Proceed as described above. This pesto is great on bruschetta with artichokes in olive oil.

# BÉCHAMEL SAUCE

**Makes 500 ml**

**Prep: 5 minutes**

**Infuse: 15 minutes**

**Cook: 10 minutes**

**1 small onion**

**1 clove**

**12 black peppercorns**

**1 bay leaf**

**1 blade of mace**

**4 parsley stalks**

**500 ml milk**

**25 g butter**

**25 g flour**

**salt and freshly ground white pepper**

Slice the onion into a saucepan with the slicing drum on the rotor vegetable slicer/shredder on speed 4. Add the clove, peppercorns, bay leaf, blade of mace, parsley stalks and milk. Bring to the boil over a gentle heat, then immediately remove from the heat, cover and infuse for 15 minutes.

Strain the milk and discard the flavourings. Melt the butter in a heavy-based saucepan, sprinkle in the flour and stir to mix. Cook over a gentle heat for 2 minutes, stirring constantly. Gradually add the milk and keep stirring until the sauce is smooth. Cook the béchamel for 5 minutes over a low heat to cook out the flour and thicken the sauce. Remove from the heat and add a flavouring of your choice or use as is. Cover the béchamel with clingfilm if you are not planning to use it immediately; this will prevent a skin from forming on the surface as it cools.

## SAFFRON MASH

**Serves 4**

**Prep: 5 minutes**

**Cook: 20 minutes**

**500 g floury potatoes**

**125 g softened butter**

**2-3 capsules powdered saffron**

**salt and freshly ground white pepper**

Peel the potatoes and cut them into same-size chunks. Boil the potatoes in salted water until tender. Drain the potatoes well, then return them to the pan and place on a gentle heat to dry out. Purée the potatoes into the mixer bowl with the fine grinding plate on the food grinder on speed 4. Add the butter and saffron, and beat with the flat beater on speed 2 until well-incorporated and smooth. Season to taste and serve with white fish such as halibut or sea bass.

## SHORTCRUST PASTRY

**Makes about 300 g**

**(enough for 1 large tart or 6 small tartlets)**

**Prep: 10 minutes**

**Chill: 30 minutes**

**200 g flour**

**¼ tsp salt**

**2 tbsp icing sugar (for Sweet shortcrust pastry)**
**100 g cold butter**

23

**1 beaten egg**

**1 tsp lemon juice**

**1-2 tbsp iced water**

Place the flour, salt and icing sugar (if using) in the mixer bowl. Dice the butter and add to the bowl. Mix with the flat beater on speed 2 until the mixture resembles breadcrumbs.

Add the egg, lemon juice and iced water. Continue to knead until a smooth dough is formed. Turn out onto a lightly floured surface and knead briefly. Shape into a ball, wrap in clingfilm and chill for at least 30 minutes.

## SHORTBREAD PASTRY

**Makes about 350 g**

**(enough for 1 large tart or 6 small tartlets)**

**Prep: 10 minutes**

**Chill: 2 hours**

**100 g softened butter**

**75 g icing sugar**

**1 egg yolk**

**½ tsp vanilla extract**

**150 g flour**

Beat the butter and icing sugar in the mixer bowl with the flat beater on speed 2 until creamy and well-blended. Add the egg yolk and vanilla and mix for 10 seconds. Sieve the flour and mix in on speed 1. Turn out onto a floured surface and shape into a ball. Wrap the pastry in clingfilm and chill for at least 2 hours.

## LEMON AND FENNEL SHORTBREAD

**Makes 24-30**

**Prep: 10 minutes**

**Rest: 1 hour**

**Cook: 15 minutes**

**225 g softened butter**

**grated zest of 2 lemons**

**115 g sugar**

**1 tsp fennel seeds**

**225 g flour**

**115 g cornflour**

Preheat the oven to 170°C/gas mark 3. Place the butter and lemon zest in the mixer bowl. Blitz the sugar and fennel seeds in a coffee mill. Add to the mixer bowl and mix with the flat beater on speed 6 until the mixture is pale and fluffy.

Sieve the flour and cornflour into the mixer bowl. Mix in with the flat beater on speed 2 until you obtain a smooth dough. Turn out and form into a ball. Wrap in clingfilm and chill for 1 hour, or until the dough is firm enough to roll out.

Roll out the dough on a lightly floured surface and stamp out shapes with acutter of your choice. Place these on a baking sheet lined with greaseproof paper. Bake for 15 minutes until the shortbread is light golden in colour. Leave to cool on a wire rack.

**Variations**

**Poppy seed shortbread**

Replace the lemon zest and fennel seeds with 1 tablespoon of poppy seeds.

**Stem ginger shortbread**

Replace the lemon zest and fennel seeds with 50 g chopped stem ginger.

# FLAKY PASTRY

**Makes about 400 g**

**Prep: 20 minutes**

**Chill: 2 hours**

**225 g flour**

**a pinch of salt**

**150 g butter**

**approx 100 ml chilled water**

Sieve the flour and salt into the mixer bowl. Dice 45 g butter and scatter over the flour. Blend with the flat beater on speed 2 until the mixture resembles breadcrumbs. Gradually mix the water into the mixture on speed 1 until you obtain a soft dough. Wrap in clingfilm and chill for 30 minutes.

Roll the dough on a lightly floured surface into a 30 cm x 10 cm rectangle. Cut 35 g butter into very small dice and and dot these evenly over the top two thirds of the pastry. Leave a good margin around the edges. Fold the bottom third of the pastry up and over the butter, then fold the top third down over the whole. Press the two open sides of the parcel together with a rolling pin to seal in the butter. Wrap in clingfilm and chill for 10 minutes.

Give the dough a half turn. Repeat the rolling and folding process once more without adding any butter, then wrap and chill for 10 minutes. Roll and fold the dough twice more, using 35 g finely diced butter butter each time. Wrap and chill for 10 minutes. Roll and fold the dough one final time without adding any butter. Wrap and chill for at least 1 hour before using.

# CHOUX PASTRY

## Makes 300 g

## (enough for 30 choux or 15 éclairs)

## Prep: 15 minutes

**150 g flour**

**1 tsp sugar**

**½ tsp salt**

**100 g butter**

**125 ml milk**

**125 ml water**

**4 eggs**

Sieve the flour, sugar and salt into a bowl or onto a sheet of greaseproof paper. Gently heat the butter and milk with the water until the butter has melted. Raise the heat and, as soon as the liquid is boiling fast, tip in all the flour in one go. Beat vigorously to form a smooth paste. Continue to beat on a low heat until the mixture forms a ball and leaves the sides of the pan. This will take about 30 seconds.

Remove from the heat and transfer the contents of the pan to the mixer bowl. Leave to cool for 2 minutes. Gradually beat in the eggs with the flat beater on speed 4 until you obtain a shiny dough; you may not need all the egg. The pastry is ready when it drops reluctantly from a wooden spoon.

## STRUDEL PASTRY

**Makes enough for 1 strudel**

**Prep: 10 minutes**

**Rest: 30 minutes**

**300 g flour**

**1 tsp salt**

**2 egg yolks**

**3 tbsp sunflower oil**

**125 ml water**

Sieve the flour and salt into the mixer bowl and mix with the dough hook on speed 1. Mix the egg yolks and sunflower oil with the water. Add to the flour with the dough hook on speed 1. Knead to a soft and sticky dough. Add up to 2 tablespoons of water, if necessary. Continue to knead on speed 2 for 4 to 5 minutes until the dough is smooth and elastic. Cover and leave to rest for 30 minutes.

## VANILLA CUSTARD

**Makes 350 ml**

**Prep: 10 minutes**

**Infuse: 20 minutes**

**Cook: 15 minutes**

**250 ml milk**

**250 ml double cream**

**1 vanilla pod**

**5 egg yolks**

**50 g sugar**

Bring the milk and cream to the boil with the split vanilla pod. Remove from the heat and leave to infuse for 20 minutes. Beat the egg yolks and sugar in the mixer bowl with the wire whisk on speed 6 for 2 minutes until very pale and thick.

Bring the milk and cream back to the boil, remove the vanilla pod and slowly pour into the creamed mixture. Mix with the wire whisk on speed 1 until amalgamated. Pour the custard back into the saucepan and cook on a low heat, stirring continuously, until the custard thickens and coats the back of a wooden spoon. Serve hot or cold. If serving cold, pour the custard into a bowl and cover the surface with clingfilm to prevent a skin from forming on the surface as it cools.

## Variations

### Bay leaf and orange custard

Infuse the milk and cream with the zest of

1 orange and 2 fresh bay leaves instead of the vanilla pod. Continue as described above.

### Saffron and orange flower custard

Infuse the milk and cream with a good pinch of saffron threads instead of the vanilla pod, then continue as described above. Stir in orange flower water to taste at the end.

# PASTRY CREAM

**Makes 500 ml**

**Prep: 10 minutes**

**Infuse: 20 minutes**

**Cook: 5 minutes**

**500 ml milk**

**1 vanilla pod**

**6 egg yolks**

**125 g sugar**

**40 g flour or cornflour**

Bring the milk to the boil with the split vanilla pod. Remove from the heat and leave to infuse for 20 minutes. Beat the egg yolks and sugar in the mixer bowl with the wire whisk on speed 6 for 2 minutes until very pale and thick. Sieve over the flour or cornflour and beat until well-blended.

Bring the milk back to the boil, remove the vanilla pod and slowly pour onto the creamed mixture. Mix with the wire whisk on speed 1 until amalgamated. Pour the cream back into the saucepan and cook on a low heat, stirring with a wooden spoon until the cream comes to the boil and starts to thicken. Gently simmer for 1 minute to cook out the flour, stirring all the time. Remove from the heat, transfer to a bowl and cover the surface with clingfilm to prevent a skin from forming on the surface as it cools.

## REAL VANILLA ICE CREAM

**Serves 4-6**

**Prep: 5 minutes**

**Infuse: 30 minutes**

**Cook: 15 minutes**

**Freeze: overnight**

**1 vanilla pod**

**300 ml milk**

**4 large egg yolks**

**100 g sugar**

# 300 ml double cream

Split the vanilla pod and scrape out the seeds. Put both the seeds and the pod in a saucepan. Add the milk and bring to the boil. Remove from the heat immediately, cover and leave to infuse for 30 minutes.

Whisk the egg yolks and sugar in the mixer bowl with the wire whisk on speed 6 until pale and thick. Bring the infused milk back to the boil, then pour onto the egg yolks and whisk on speed 4.

Rinse out the saucepan and pour the vanilla cream back into the pan. Stir on a gentle heat until the custard thickens and coats the back of a spoon. Strain into a bowl and cover the surface with clingfilm to prevent a skin from forming on the surface. Chill.

When the custard is completely cold, stir in the double cream and pour into the freeze bowl.

Churn with the dasher on speed 2 until almost firm. Spoon into a freezer-proof container and freeze overnight. The next day, place in the refrigerator for 20 minutes before serving to soften.

## Variations

### Caramel ice cream

Double the amount of sugar and make a caramel with 75 ml water. Infuse the milk with the vanilla but leave the pod whole. Pour onto the caramel but be careful: the mixture will spit and solidify. Heat gently until the caramel is smooth again, then continue with the recipe as described above.

### White chocolate ice cream

Make the custard as described above but use 6 egg yolks instead of 4. Then pour the custard onto 125 g grated white chocolate and stir until smooth. Strain, then chill and continue as described above.

## CHOCOLATE GANACHE

**Makes 250 ml**

**Prep: 2 minutes**

**Cook: 5 minutes**

**150 g dark chocolate (70%)**

**100 ml double cream**

Grate the chocolate with the coarse shredding drum on the rotor vegetable slicer/shredder on speed 4. Place in a heatproof bowl. Bring the cream to the boil on a low heat. Pour onto the chocolate and leave for 2 minutes, then stir gently until the chocolate has melted. Leave to cool.

**Variations**

**Milk chocolate and nutmeg ganache**

Replace the dark chocolate with 165 g milk chocolate and add ¼ to ½ teaspoon freshly grated nutmeg.

**White chocolate and rose liqueur ganache**

Replace the dark chocolate with 175 g white chocolate, and add 1 to 2 tablespoons of rose liqueur when the ganache is still warm but not hot.

# GIN AND TONIC SYLLABUB

**Serves 4**

**Rest: 2 hours**

**Cook: 5 minutes**

**juice of 1 lemon**

**grated zest of 2 lemons**

**2 tbsp good-quality gin**

**50 g sugar**

**200 ml double cream**

**100 ml sweet sparkling wine**

Place the lemon juice and zest, gin and sugar in a bowl and stir to dissolve the sugar. Leave for at least 2 hours.

Whisk the cream in the mixer bowl with the wire whisk on speed 6 until it just holds its shape. Gradually whisk in the sparkling wine on speed 4; do not add the wine too fast or the cream will split. Reduce to speed 2 and add the lemon and sugar mixture. Keep whisking until the syllabub is thick and fluffy.

# Mont Blanc macaroons

**Makes 40 macaroons**

**Prep: 10 minutes**

**Rest: 15 minutes**

**Cook: 25 minutes**

**1 vanilla pod**

**4 egg whites**

**a pinch of cream of tartar**

**25 g sugar**

**225 g icing sugar**

**125 g ground almonds**

**200 g crème de marrons (sweetened chestnut purée), for filling**

Preheat the oven to 150°C/gas mark 2. Line two baking sheets with greaseproof paper or silicone sheets. Split the vanilla pod and scrape the seeds into the mixer bowl. Add the egg whites and whisk with the wire whisk on speed 4 until frothy.

Increase to speed 8 and beat in the cream of tartar until soft peaks form. Reduce to speed 4 and add the sugar, then increase to speed 8 again and whisk until the meringue is firm.

Sieve the icing sugar and ground almonds twice, then carefully fold into the meringue with a large metal spoon. The batter should be smooth and shiny. Fill a piping bag with a 1 cm plain nozzle and pipe little heaps of batter 2 cm in diameter onto the baking sheets. Leave 2.5 cm between the macaroons, as they spread during baking. You could also use a teaspoon to drop heaps of batter onto the baking sheets. You should have 80 little heaps, enough for 40 filled macaroons. Leave to rest for 15 minutes.

Bake the macaroons for 20 to 25 minutes. The macaroons are ready when they have risen and feel dry and firm to the touch. Allow to cool for 2 to 3 minutes before removing the macaroons from the greaseproof paper. Leave to cool completely before sandwiching with chestnut purée.

# 3.

# STARTERS

# TARAMASALATA

Homemade taramasalata is so much tastier and more appealing than the bubblegum-pink shop-bought version. And it is a doddle to make too.

**Serves 8**

**Prep: 15 minutes**

**Chill: 1 hour**

**50 g sliced white bread**

**4 tbsp warm water**

**200 g tarama (smoked cod's roe)**

**150 ml olive oil**

**juice of 1 lemon**

**finely chopped chives, to garnish (optional)**

**salt and freshly ground black pepper**

Remove the crusts from the bread and tear the bread into pieces. Pour over the water and leave for 10 minutes. Place the bread in the blender and process briefly on mix speed . Add the tarama and blend again until a paste is formed.

With the motor running, gradually add the olive oil and lemon juice until you obtain a smooth purée. Add a tablespoon or two of hot water if the taramasalata is too thick. Scrape into a bowl and season to taste; do not add too much salt. Cover and chill for 1 hour. Just before serving, give the taramasalata a quick stir and sprinkle over the chives, if desired. Serve with pitta bread.

## AUBERGINE CAVIAR

**Serves 8**

**Prep: 5 minutes**

**Cook: 1 hour**

**500 g aubergines**

**2 crushed garlic cloves**

**1 tsp ras el hanout (Moroccan spice mix)**

**2 tbsp olive oil**

**50 g shallots**

**½ tsp harissa**

**50 ml extra virgin olive oil**

**2 tbsp lemon juice**

**100 g peeled, deseeded and diced plum tomatoes**

**1 tbsp chopped coriander**

**salt and freshly ground black pepper**


Preheat the oven to 200°C/gas mark 6. Halve the aubergines lengthways and score the flesh in a diamond pattern. Spread the garlic over the flesh, then sprinkle with the ras el hanout and 1 tablespoon of olive oil. Sandwich the aubergine halves together and wrap in aluminium foil. Bake for 45 minutes to 1 hour, or until the aubergines feel very soft to the touch.

Meanwhile, chop the shallots with the medium shredding drum on the rotor vegetable slicer/shredder on speed 4. Heat the remaining olive oil in a pan and sauté the shallots for 5 to 10 minutes. Stir in the harissa and remove from the heat.

Unwrap the aubergines and scrape the flesh into the blender. Add the shallots and process on purée speed until smooth. With the motor running, drizzle in the extra virgin olive oil until the purée is very smooth and silky. Add the lemon juice.

Spoon the aubergine caviar into a bowl and stir in the diced tomatoes and coriander. Season to taste and serve with toasted pitta bread.

## ICED GUACAMOLE WITH CHERRY TOMATO SALSA

**Serves 4-6**

**Prep: 30 minutes**

**Freeze: overnight**

**350 g ripe avocado flesh**

**1 red bird's-eye chilli**

**1 garlic clove**

**juice of 2 limes**

**grated zest of 1 lime**

**2 tbsp finely chopped coriander**

**½ tsp salt**

**1 tbsp sugar**

**200 ml double cream**

**ready-salted tortilla chips, to serve**

**Salsa:**

**4 spring onions**

**½ red onion**

**12 cherry tomatoes**

**1 tbsp olive oil**

**juice of ½ lime**

**1 tbsp finely chopped coriander**

**salt and freshly ground black pepper**

Place the avocado flesh in the blender. Deseed the chilli and add to the blender, together with the garlic, lime juice and zest, coriander, salt, sugar and cream.

Process at purée speed until the mixture is well-blended but flecks of chilli and coriander are still visible. Place the mixture in the freeze bowl and churn with the dasher on speed 1 until almost firm. Spoon into a freezer-proof container and freeze overnight.

The next day, place the ice cream in the refrigerator for 30 minutes before serving to soften. Meanwhile, chop the spring onions, red onion and cherry tomatoes. Mix with the olive oil, lime juice and coriander. Season to taste. Place the ice cream in cocktail glasses and spoon over the cherry tomato salsa. Serve with tortilla chips.

## TZATZIKI WITH CUCUMBER GRANITA

**Serves 6-8**

**Prep: 25 minutes**

**Chill: 10 minutes**

**Freeze: 4 hours**

**2 cucumbers**

**1 tsp salt**

**1 tsp dried oregano**

**450 g Greek yoghurt**

**2 crushed garlic cloves**

**1 tbsp white wine vinegar**

**2 tbsp extra virgin olive oil**

**3 tbsp finely chopped mint**

**freshly ground black pepper**

**Cucumber granita (optional):**

**50 g sugar**

47

**200 ml water**

**1 tbsp finely chopped mint**

**1 kg cucumbers**

**1 tbsp lemon juice**

**a pinch of salt**

**pitta crisps, to serve**

This Greek dip is lovely on its own with bread or as part of a mezze platter with Taramasalata and Aubergine caviar. For a more elegant presentation, serve it in glasses with cucumber granita and pitta crisps.

Shred the cucumbers into the mixer bowl with the medium shredding drum on the rotor vegetable slicer/shredder on speed 4. Sprinkle with the salt and oregano, and leave to drain for 10 minutes. Pour off the excess liquid, then place the cucumber in a sieve and gently squeeze out any remaining liquid.

Return the cucumber to the mixer bowl and stir in the yoghurt with the flat beater on speed 2. Add the garlic, vinegar, olive oil and mint. Stir well and season to taste. Chill for 10 minutes, then serve.

Make the cucumber granita, if using. Put the sugar in a pan with the water. Slowly bring to the boil, stirring until the sugar has dissolved. Add the mint, then simmer on a low heat for 2 to 3 minutes. Peel and deseed the cucumbers, then process in the blender at liquefy speed. Sieve into a freezerproof container, then stir in the mint syrup, lemon juice and salt. Chill first, then freeze for 2 hours until the granita is firm around the edges.

Break up the ice crystals with a fork and stir them into the granita. Return to the freezer for 30 minutes and fork through again. Repeat this freezing and forking through until the granita is fluffy, then scrape onto the tzatziki. Serve with pitta crisps.

## FOAM OF HUMMUS

### Serves 6

### Prep: 15 minutes

### 400 g tinned chickpeas
### 1 garlic clove
### 150 ml double cream

**1 tbsp tahini (sesame seed paste)**

**20 g olive oil**

**½ tsp salt**

**¼ tsp paprika**

**½ tsp ground cumin**

Drain the chickpeas but reserve their liquid. Rinse the chickpeas under cold running water, then purée with the garlic into the mixer bowl with the fruit and vegetable strainer on speed 4.

Bring the cream to the boil with 50 ml of the reserved chickpea liquid. Mix this, as well as the tahini, olive oil, salt, paprika and cumin into the chickpea purée with the wire whisk on speed 4. Pour into a 500 ml siphon and screw on two $N_2O$ capsules. Shake well, then pipe into glasses and serve at once.

## Hummus bi tahina: classic recipe

Drain the chickpeas, discarding their liquid, and rinse them under cold running water. Purée the chickpeas and garlic as above, adding the tahini, salt, paprika and cumin as well. Place the flat beater and omit the double cream but increase the olive oil to 100 ml. Mix for 1 minute on speed 4. Stir in the juice of 1 lemon and a handful of finely chopped coriander. Serve with toasted pitta bread.

# ICED COURGETTE AND BASIL SOUP WITH LEMON OIL

**Serves 4**

**Prep: 5 minutes**

**Cook: 20 minutes**

**Chill: overnight**

**2 courgettes**

**2 tbsp olive oil**

**1 bunch of basil**

**2 tbsp lemon olive oil, plus extra for drizzling**

**50 g toasted pine nuts**

**4 ice cubes**

**25 g salmon roe, to garnish**

**salt and freshly ground black pepper**

Slice the courgettes with the slicing drum on the rotor vegetable slicer/shredder on speed 4. Heat the olive oil in a large saucepan and sauté the courgettes for 5 minutes. Cover with water, bring to the boil and cook for 5 minutes until tender. Leave to cool for 15 minutes.

Pour the contents of the saucepan into the blender with the basil, lemon oil and pine nuts. Mix until smooth; you may need to do this in batches. Season to taste and chill. Pour the soup into chilled bowls or glasses and drop an ice cube into each portion. Garnish with salmon roe and a drizzle of lemon oil.

## GAZPACHO

**Serves 4-6**

**Prep: 15 minutes**

**Chill: overnight**

**750 g ripe plum tomatoes**

**1 red pepper**

**1 red onion**

**½ cucumber**

**2 garlic cloves**

**100 ml tomato passata**

**a few drops of Tabasco**

**4 tbsp olive oil**

**2 tbsp balsamic vinegar**

**100 ml cold water**

**2 hard-boiled eggs**

**8 spring onions**

**12 gherkins**

**½ yellow pepper**

**100 g finely diced cooked ham**

**extra virgin olive oil, to serve**

**salt and freshly ground black pepper**

Roughly chop the tomatoes, pepper, onion and cucumber. Purée the vegetables and garlic into the mixer bowl with the coarse grinding plate on the food grinder on speed 4. Add the passata, Tabasco, olive oil, balsamic vinegar and water. Mix thoroughly with the flat beater on speed 4. Season to taste. Refrigerate overnight for the soup to chill thoroughly and the flavours to develop.

The next day, finely chop the eggs, ham, spring onions, gherkins and yellow pepper. Pour the gazpacho into chilled bowls and drizzle some extra virgin olive oil into each bowl. Sprinkle the garnishes over the soup.

## RUSTIC PORK TERRINE WITH MUSHROOMS AND FRESH HERBS

**Serves 8**

**Prep: 20 minutes**

**Chill: 1-2 days**

**Cook: 2 hours**

**3 shallots**

**3 garlic cloves**

**125 g mushrooms**

**3 tbsp butter**

**450 g pork belly in one piece, rind and bones removed**

**175 g pork fillet**

**175 g chicken livers**

**150 g pancetta, in one piece**

**2 tbsp finely chopped flat-leaf parsley**

**1 tbsp finely chopped sage**

**1 tbsp finely chopped thyme**

**2 tbsp finely chopped basil**

**2-3 tbsp Vecchia Romagna (Italian brandy)**

**10-12 rindless back bacon rashers**

**Plum, pear and sultana chutney, to serve**

**salt and freshly ground black pepper**

Making your own terrine requires a little work but it is well worth the effort. Instead of the chutney, you could serve this flavoursome terrine with mostarda di frutta, fruit preserved in a mustard syrup.

Preheat the oven to 170°C/gas mark 3. Chop the shallots, garlic and mushrooms with the coarse shredding drum on the rotor vegetable slicer/shredder on speed 4. Melt the butter in a frying pan and sauté the shallots, garlic and mushrooms for 10 minutes on a high heat. Cool.

Mince the pork belly, pork fillet, chicken livers and pancetta into the mixer bowl with the fine grinding plate on the food grinder on speed 4. Change to the flat beater, then mix in the mushroom mixture, herbs and Vecchia Romagna on speed 2. Season to taste.

Line a 1-litre terrine dish with the bacon rashers, letting the ends hang over the sides. Spoon the meat mixture into the dish and press down firmly. Fold the bacon over the top and cover with either a lid or a double sheet of aluminium foil.

Place in a roasting tin half-filled with hot water and cook in the oven for 1 hour 45 minutes to 2 hours until the terrine shrinks away from the sides of the dish and a skewer inserted in the centre comes out clean.

Remove the terrine from the roasting tin and cool for 1 hour. Remove the aluminium foil or lid, cover with clean foil and weigh down with several tins. Cool completely, then chill with the weights for 1 to 2 days. Return the terrine to room temperature before serving. Cut into slices and serve with the chutney. Eat within 2 to 3 days.

# GARLIC MUSHROOM BRUSCHETTA

**Serves 6**

**Prep: 5 minutes**

**Cook: 20 minutes**

**6 shallots**

**4 garlic cloves**

**450 g mushrooms**

**3 tbsp olive oil**

**100 ml dry white wine**

**3 tbsp finely chopped rosemary**

**6 tbsp mascarpone**

**12 ciabatta slices**

**Parmesan shavings, to garnish**

**salt and freshly ground black pepper**

Chop the shallots and 3 garlic cloves with the coarse shredding drum on the rotor vegetable slicer/shredder on speed 4. Slice the mushrooms with the slicing drum. Heat the olive oil in a large frying pan and sauté the shallots and garlic for 5 minutes until softened. Add the mushrooms and sauté for 10 minutes on a high heat until golden brown.

Add the wine and reduce. Stir in the mascarpone and rosemary, then season to taste and remove from the heat.

Toast the ciabatta slices and halve the last garlic clove. Rub the cut ends of the garlic over one side of each piece of toast. Top the bread with the mushroom mixture and garnish with a few Parmesan shavings. Serve immediately.

## CHOCOLATE TAPENADE CROSTINI

**Serves 6**

**Prep: 10 minutes**

**60 g dark chocolate (70%)**

**300 g stoned black olives**

**1 sprig of thyme**

**5 tbsp olive oil**

**24 baguette slices**

**100 g mature goat's cheese**

**freshly ground black pepper**

Melt the chocolate. Mix the olives and thyme leaves in the blender . With the motor running, gradually pour in the olive oil and melted chocolate until the mixture has a spreadable consistency. Season with pepper and spoon into a bowl, cover and chill. Lightly toast the baguette slices on both sides until crispy, then top with the tapenade. Crumble over the goat's cheese and serve immediately.

## BLINIS WITH SOURED CREAM AND BEETROOT-CURED SALMON

**Serves 6**

**Prep: 40 minutes Rest: overnight Rise: 1 hour**

**Cook: 35 minutes**

**100 g buckwheat flour**

**100 g flour**

**7 g dried yeast**

**1 tbsp sugar**

**½ tsp salt**

**300 ml warm milk**

**2 eggs**

**50 g melted butter, for frying**

**Beetroot-cured salmon:**

**400 g skinless salmon fillet**

**1 ½ tbsp sea salt**

**2 tsp sugar**

**3 tbsp chopped dill**

**75 ml vodka**

**125 g cooked beetroot**

**200 g soured cream**

**caviar or black lumpfish roe, to garnish**

Start this recipe the day before by making the beetroot-cured salmon. Rinse the salmon and pat dry with kitchen paper. Mix the salt, sugar and dill and rub all over the salmon. Place the salmon in a shallow dish and pour over the vodka. Cover and chill for at least 4 hours or overnight. Turn over once.

Next day, grate the beetroot with the medium shredding drum on the rotor vegetable slicer/shredder on speed 4. Remove the salmon from the marinade and mix the grated beetroot into the juices. Return the salmon to the marinade and chill for another 6 hours.

Place the buckwheat flour, plain flour, yeast, sugar and salt in the mixer bowl. Separate the eggs. Mix the milk and egg yolks into the dry ingredients with the wire whisk on speed 4 until smooth. Cover and leave to rise at room temperature for 1 hour, or until doubled in volume.

Transfer the batter to another bowl. Clean and dry the mixer bowl and wire whisk thoroughly. Whisk the egg whites with the wire whisk on speed 8 until stiff, then fold them into the batter.

Heat a little melted butter in a large frying pan on a medium heat. Add 4 heaps of batter, about 1 ½ tablespoons per blini, to the frying pan. Cook the blinis for 45 seconds on one side, then flip them over carefully and cook for another 30 seconds. Remove the blinis from the frying pan and keep them warm in a low oven. Continue to cook the rest of the blinis, brushing the frying pan with melted butter each time.

Rinse the salmon and pat dry, then cut the salmon into 5 mm dice. Garnish each blini with a dollop of soured cream and arrange the salmon on top. Garnish with a little caviar or lumpfish roe.

## CELERIAC, APPLE AND PECAN RÉMOULADE

**Serves 6**

**Prep: 15 minutes**

**5 tbsp Wasabi mayo**

**5 tbsp crème fraîche**

**1 tbsp lemon juice**

**600 g celeriac**

**2 red apples**

**2 tbsp finely chopped flat-leaf parsley**

**50 g chopped pecan nuts**

**salt and freshly ground white pepper**

Place the wasabi mayo, crème fraîche and lemon juice in the mixer bowl and whisk together with the wire whisk on speed 6. Season to taste.

Peel the celeriac and cut into chunks. Grate the celeriac into the mixer bowl with the medium shredding drum on the rotor vegetable slicer/shredder on speed 4. Toss the grated celeriac into the dressing immediately. Core and grate the apples into the bowl with the coarse shredding drum. Stir in the parsley and pecan nuts and serve.

## TRUFFLED VITELLO TONNATO

**Serves 6**

**Prep: 25 minutes**

**Marinate: overnight**

**Cook: 1 hour**

**1 onion**

**1 carrot**

**1 celery stalk**

**1 lemon**

**600 g veal loin, in one piece**

**1 bay leaf**

**2 cloves**

**½ l dry white wine, preferably Italian**

**1 tsp sea salt**

**½ l water**

**Dressing:**

**200 g drained tinned tuna**

**4 anchovy fillets**

**2 tbsp capers**

**2 egg yolks**

**juice of ½ lemon**

**250 ml olive oil**

**2 tbsp Crema Tartufata (white truffle cream)**

**a handful of caperberries, to garnish**

**freshly ground black pepper**

For an even more luxurious version of this dish, you could replace the truffle cream with 15 g finely chopped fresh black truffle.

Slice the onion, carrot, celery and lemon into a large bowl with the slicing drum on the rotor vegetable slicer/shredder on speed 4. Add the veal, bay leaf and cloves. Pour over the white wine, cover and chill overnight. Regularly turn the meat over while it is marinating.

The next day, put the contents of the bowl into a large pan. Add the salt and water. Bring to the boil, then reduce the heat and cover. Simmer for 45 minutes to 1 hour until the veal is cooked and tender. Leave to cool in the pan.

Make the dressing. Place the first 5 ingredients in the blender and process on mix speed. Gradually pour in the olive oil on speed 6 until the dressing is thick and shiny. Add about 2 tablespoons of poaching liquid to thin the dressing. Finally, mix the Crema Tartufata into the dressing on stir speed. Season with extra lemon juice, if you like.

Drain the veal from the stock and slice as thinly as possible. Arrange on serving dishes and spoon over the dressing. Garnish with a few caperberries and serve.

# NEW POTATO SALAD WITH SALSA VERDE, ROAST BEETROOT AND SMOKED EEL

Serves 4

Prep: 15 minutes

Cook: 20 minutes

250 g small beetroot

2 tbsp olive oil

500 g new potatoes

200 g smoked eel

salt and freshly ground black pepper

Salsa verde:

1 garlic clove

2 tbsp capers

2 tbsp gherkins

3 anchovy fillets

½ tbsp Dijon mustard

a handful of flat-leaf parsley

1 bunch of basil

1 bunch of mint

125 ml extra virgin olive oil

**1 tbsp red wine vinegar**

Preheat the oven to 200°C/gas mark 6. Peel the beetroot and cut into wedges. Toss with the olive oil, season and place in a roasting tin. Roast for 30 minutes until tender.

Prepare the salsa verde. Place the first 5 ingredients in the blender. Pick the leaves from the herbs and add to the blender. Process on stir speed to a coarse purée. While the motor is running, drizzle in the olive oil until fully incorporated. Add the vinegar and season.

Cut the potatoes into same-size chunks, if necessary, and boil in salted water until tender to the bite. Drain and immediately mix in two thirds of the salsa verde. Mix the remaining salsa verde with the beetroot. Spoon onto plates and arrange the smoked eel on top.

# RED CABBAGE, BEETROOT AND CRANBERRY SALAD

**Serves 6**

**Prep: 20 minutes**

**Chill: overnight**

**½ small red cabbage**

**1 medium beetroot**

**1 small red onion**

**50 g walnuts**

**50 g dried cranberries**

**Dressing:**

**1 tbsp lemon juice**

**4 tbsp red wine vinegar**

**1 tbsp honey**

**4 tbsp walnut oil**

**6 tbsp olive oil**

**½ tsp caraway seeds**

**salt and freshly ground black pepper**

Remove the outer leaves and central core of the red cabbage. Slice the cabbage into a colander with the slicing drum on the rotor vegetable slicer/shredder on speed 4. Rinse well under cold running water until the water runs clear. Drain thoroughly, then place the red cabbage in a bowl.

Peel the beetroot and red onion. Grate both into the bowl containing the red cabbage with the coarse shredding drum on the rotor vegetable slicer/shredder on speed 4. Toast and roughly chop the walnuts. Add to the cabbage salad with the cranberries.

Make the dressing. Place all the ingredients in the mixer bowl and blend with the wire whisk on speed 4 until smooth. Pour over the cabbage salad, toss lightly and chill overnight before serving.

# COURGETTE SALAD WITH MINT, RED CHILLI AND LEMON

**Serves 4**

**Prep: 5 minutes**

**Cook: 10 minutes**

**Marinate: 15 minutes**

**4 small courgettes**

**1 mild red chilli**

**1 garlic clove**

**2 tbsp olive oil**

**4 tbsp extra virgin olive oil**

**juice of ½ lemon**

**1 bunch of mint**

**100 g feta**

**salt and freshly ground black pepper**

Slice the courgettes, chilli and garlic with the slicing drum on the rotor vegetable slicer/shredder on speed 4. Brush the courgettes with olive oil and cook on a griddle pan until golden brown and just tender. Mix the extra virgin olive oil with the lemon juice, then finely chop the mint and stir into the vinaigrette. Add the sliced chilli and garlic, season to taste and pour the vinaigrette over the courgettes. Marinate for 15 minutes, then crumble over the feta and serve.

# CHORIZO, SAFFRON AND PARSLEY TORTILLA

**Serves 6-8**

**Prep: 15 minutes**

**Rest: 15 minutes**

**Cook: 15 minutes**

**150 g spicy chorizo sausage**

**600 g peeled potatoes**

**1 large onion**

**150 ml extra virgin olive oil**

**6 eggs**

**2 capsules powdered saffron**

**3 tbsp finely chopped flat-leaf parsley**

**salt and freshly ground black pepper**

Peel and slice the chorizo into a bowl with the slicing drum on the rotor vegetable slicer/shredder on speed 4. Slice the potatoes and onion as well. Heat 1 tablespoon of olive oil in a large deep frying pan and sauté the chorizo until golden brown. Remove from the pan with a slotted spoon and leave to drain on kitchen paper. Pour away the oil and wipe out the pan.

Heat 100 ml olive oil in the frying pan and sauté the potatoes and onions for 10 to 15 minutes until the vegetables are golden brown and tender. Remove from the heat and cool for 15 minutes.

Beat the eggs, saffron and parsley in the mixer bowl with the wire whisk on speed 2. Stir in the chorizo, and potatoes and onions with their oil. Season to taste. Heat the remaining olive oil in the frying pan and pour in the egg mixture. Cook on a low heat for 10 minutes until the tortilla is almost cooked through.

Carefully slide the tortilla onto a large plate and place a large lid on top. Invert the tortilla onto the lid and slide back into the pan. Cook for another 5 minutes, then cool to room temperature before serving.

# ASIAN COLESLAW

**Serves 6**

**Prep: 15 minutes**

**Chill: overnight**

**500 g white cabbage**

**100 g carrots**

**75 g shallots**

**2 tbsp finely chopped coriander**

**2 tbsp finely chopped mint**

**50 g roasted salted peanuts**

**salt and freshly ground black pepper**

**Dressing:**

**50 g sugar**

**75 ml rice wine vinegar**

**juice of 1 lime**

**2 tbsp sweet chilli sauce**

**2 tbsp fish sauce**

**1 crushed garlic clove**

Prepare the dressing first. Put the sugar and vinegar in a pan. Bring slowly to the boil, stirring until the sugar has dissolved. Remove from the heat and add the lime juice, chilli and fish sauces, and garlic.

Remove the outer leaves and hard inner core of the cabbage and cut the cabbage into wedges. Peel the carrots and halve the shallots. Shred the cabbage, carrots and shallots into the mixer bowl with the coarse shredding drum on the rotor vegetable slicer/shredder on speed 4. Pour over the dressing, season to taste and mix well. Cover and chill overnight. Stir the herbs into the coleslaw just before serving and sprinkle the roughly chopped peanuts on top.

## MINESTRONE WITH GREMOLATA

### Serves 4-6

### Prep: 20 minutes + overnight soaking if using dried beans

### Cook: 40 minutes

### 400 g tinned borlotti beans, or 200 g dried borlotti beans

2 carrots

1 small leek

2 small courgettes

1 onion

2 garlic cloves

4 tbsp olive oil

4 sprigs of rosemary

400 g tinned chopped tomatoes

4 tbsp tomato purée,

preferably of sun-dried tomatoes

1.5 l water

100 g spaghetti

freshly grated Parmesan cheese, to serve

salt and freshly ground black pepper


Gremolata:

2 garlic cloves

½ bunch of flat-leaf parsley grated

zest of 1 lemon

Served with some crusty bread or fragrant garlic bread, minestrone is a meal in itself. Replace the borlotti beans with cannellini beans, if you prefer.

If using dried beans, rinse them well and cover with plenty of cold water.

Soak overnight. The next day, drain the beans well, then place them in a saucepan and cover with fresh water. Bring to the boil, then drain and cover again with cold water. Return to the boil and simmer for 15 minutes, then drain and reserve the beans. If using tinned beans, drain and rinse them well.

Slice the vegetables and garlic into the mixer bowl with the slicing drum on the rotor vegetable slicer/shredder on speed 4. Heat the olive oil in a large saucepan, then add the vegetables and sauté for 5 minutes.

Finely chop the rosemary, then add half to the saucepan and cook for another minute. Finally, add the tinned tomatoes and tomato purée, cover with the water and bring to the boil. Simmer for 20 minutes, or until the vegetables are al dente.

Break up the spaghetti into smallish pieces and add to the pan with the borlotti beans. Cook for another 5 to

10 minutes until the pasta is al dente. Stir in the remaining rosemary and remove the saucepan from the heat. Season to taste and leave for a few minutes. Meanwhile, make the gremolata. Finely chop the garlic and parsley, then mix in the lemon zest. Ladle the minestrone into bowls and sprinkle over the gremolata. Serve with Parmesan cheese.

## PEA, MARJORAM AND MASCARPONE SOUP

**Serves 6**

**Prep: 10 minutes**

**Cook: 25 minutes**

**2 large shallots**

**2 garlic cloves**

**50 g butter**

**800 g frozen peas**

**1 l chicken or vegetable stock**

**125 g mascarpone**

**3 tbsp chopped marjoram**

**6 pancetta slices**

**3 slices of white bread**

**6 fat king scallops**

**1 tbsp olive oil**

**salt and freshly ground black pepper**

Chop the shallots and garlic with the medium shredding drum on the rotor vegetable slicer/shredder on speed 4. Heat half the butter in a large saucepan and sauté the shallots and garlic for 5 minutes until softened. Stir in the peas and cook for 1 minute.

Add the stock, bring to the boil and cook for 5 minutes until the peas are tender. Leave to cool for 10 minutes. Pour the contents of the saucepan into the blender with the mascarpone and marjoram. Liquefy until smooth; you may need to do this in batches. Season to taste and keep warm.

Heat the remaining butter in a frying pan and fry the pancetta until crisp, then drain on kitchen paper. Cut the crusts off the bread and cut the bread into 1 cm cubes. Fry the bread in the pancetta butter until golden brown. Slice each scallop horizontally into three thin slices. Heat the olive oil in a frying pan and sear the scallops for 10 to 15 seconds on each side. Pour the soup into bowls and sprinkle over the croûtons. Gently place the scallops on the soup and finish with the crispy pancetta.

## PUMPKIN SOUP WITH CRISPY SAGE LEAVES AND AMARETTI CRUMBLE

**Serves 4**

**Prep: 10 minutes**

**Cook: 1 hour**

**750 g pumpkin**

**4 tbsp olive oil**

**1 red onion**

**2 garlic cloves**

**½ tsp cracked cardamom pods**

**500 ml vegetable stock**

**grated zest of 1 orange**

**a handful of sage leaves**

**salt and freshly ground black pepper**

**Amaretti crumble:**

**40 g crunchy amaretti biscuits**

**50 g ground almonds**

**¼ tsp ground cinnamon**

**20 g butter**

First prepare the amaretti crumble. Coarsely crush the amaretti biscuits and mix with the almonds and cinnamon. Melt the butter in a small pan and sauté the amaretti mixture until golden brown. Leave to cool.

Preheat the oven to 200°C/gas mark 6. Peel the pumpkin and scoop out the seeds. Cut into slices. Drizzle over 1 tablespoon of olive oil and roast in the oven for 30 to 40 minutes, or until the pumpkin is cooked through.

Chop the onion and garlic with the coarse shredding drum on the rotor vegetable slicer/shredder on speed 4. Heat 2 tablespoons of olive oil in a large saucepan and sauté the onion and garlic until softened. Cut the pumpkin flesh into cubes and add to the pan with the cardamom pods. Pour in the vegetable stock and gently simmer until the vegetables are tender.

Ladle the soup into the blender. Add the orange zest and mix on purée speed until the soup is completely smooth. Season to taste and keep warm. Heat the last tablespoon of olive oil in a small pan and fry the sage leaves until crispy. Drain on kitchen paper. Ladle the soup into bowls and sprinkle the amaretti crumble on top. Decorate with a few crispy sage leaves and serve at once.

## FRENCH ONION SOUP WITH HERBED CHEESE TOASTS

**Serves 4**

**Prep: 10 minutes**

**Cook: 40 minutes**

**750 g onions**

**4 tbsp butter**

**2 tbsp flour**

**750 ml light beef stock**

**1 tbsp finely chopped thyme**

**salt and freshly ground black pepper**

**Herbed cheese toasts:**

**60 g Gruyère cheese**

**1 tbsp finely snipped chives**

**1 tbsp finely chopped flat-leaf parsley**

**1 tsp finely chopped garlic**

**1 tbsp finely chopped shallot**

**4 tbsp melted butter**

**4 thick baguette slices**

This hearty winter soup is a French classic. For an interesting variation, use red onions and replace some of the stock with red wine. Flavour with rosemary instead of thyme.

Preheat the oven to 200°C/gas mark 6. Make the herbed cheese toasts first. Grate the cheese with the fine shredding drum on the rotor vegetable slicer/shredder on speed 4. Set aside. Stir the herbs, garlic and shallot into the melted butter. Brush the baguette slices with the herb butter and bake for 5 to 10 minutes in the oven.

Slice the onions with the slicing drum on the rotor vegetable slicer/ shredder on speed 4. Melt the butter in a large saucepan and sauté the onions in a covered pan on a low heat for 25 minutes. Dust the onions with the flour and continue to cook until golden brown, stirring constantly. Add the beef stock, mix well and cook for 10 minutes. Season to taste.

Stir the thyme into the soup and pour the soup into heatproof bowls. Divide the cheese over the toasts and place these carefully on the soup. Place the soup bowls under a preheated grill until the cheese has melted. Serve immediately.

## CAULIFLOWER AND STILTON SOUFFLÉ

**Serves 6**

**Prep: 25 minutes**

**Cook: 25 minutes**

**butter and grated Parmesan cheese, for the ramekins**

**500 g cauliflower**

**25 g butter**

**25 g flour**

**150 ml milk**

**½ tsp Espelette pepper**

**25 g cream cheese**

**3 eggs**

**75 g Stilton (or Irish Cashel Blue)**

**2 egg whites**

**salt**

Preheat the oven to 200°C/gas mark 6. Lightly butter 6 x 175 ml ramekins and dust them with grated Parmesan. Cut the cauliflower into florets and steam or cook for 8 to 10 minutes until tender. Leave to cool, then process in the blender on stir speed to an almost smooth purée. Place in the mixer bowl.

Melt the butter in a small saucepan and stir in the flour. Cook over a gentle heat for 2 minutes, then gradually mix in the milk. Heat for another 2 minutes until the sauce thickens and coats the back of a spoon. Remove from the heat, then season with salt and the Espelette pepper. Add to the mixer bowl, with the cream cheese. Mix with the flat beater on speed 2.

Separate the eggs and beat the egg yolks into the cauliflower mixture on speed 4. Clean and dry the mixer bowl thoroughly, then whisk all the egg whites with the wire whisk on speed 8 until stiff. Carefully fold into the cauliflower mixture, adding the crumbled Stilton in the process. Spoon into the prepared ramekins. Bake the soufflés for 20 to 25 minutes until golden brown and well-risen. Serve at once.

## CRAB CAKES WITH MANGO AND SWEETCORN SALSA

**Serves 4**

**Prep: 15 minutes**

**Chill: 30 minutes**

**Cook: 10 minutes**


**2 spring onions**

**2 tbsp chopped coriander**

**2 tbsp fish sauce**

**1 red bird's-eye chilli**

**2 lemongrass stalks**

**2.5 cm fresh ginger**

**1 large egg white**

**250 g fresh white crabmeat**

**1 beaten egg**

**3 tbsp dried breadcrumbs**

**3 tbsp desiccated coconut sunflower oil, for frying lime wedges, to serve**

**Salsa:**

**1 ripe but firm mango**

**2 plum tomatoes**

**150 g tinned sweetcorn**

**1 red onion**

**1 mild green chilli**

**juice of 2 limes**

**2 tbsp finely chopped coriander**

Make the salsa first. Peel and stone the mango, then cut into 5 mm dice. Deseed and cut the tomatoes in 5 mm dice. Drain the sweetcorn. Finely chop the red onion. Deseed and finely chop the chilli. Mix all the ingredients for the salsa, cover and chill.

Place the first 7 ingredients for the crab cakes in the blender and process on stir speed. Place the crabmeat, egg, breadcrumbs and coconut in the mixer bowl. Mix in the herb paste with the flat beater on speed 2. Chill for 30 minutes until firm.

Shape the crab mixture with wet hands into 8 medium-sized cakes (or 16 small ones). Heat 1 cm of sunflower oil in a heavy-based pan and cook the crab cakes for 2 minutes on each side (1 minute for the small ones) until golden brown and crispy. Serve immediately with the salsa and lime wedges.

# CARROT TARTLETS WITH SWEET AND SOUR CARROT VINAIGRETTE AND GIROLLE MUSHROOMS

**Serves 4**

**Prep: 30 minutes**

**Chill: 30 minutes**

**Cook: 45 minutes**

1 quantity Shortcrust pastry

250 g carrots

300 ml carrot juice

300 ml water

150 ml double cream

1 egg

1 egg yolk

50 ml olive oil

150 g girolle mushrooms

salt and freshly ground black pepper


Vinaigrette:

75 g finely diced carrots

175 ml olive oil

1 sprig of sage

½ tsp salt

100 ml white balsamic vinegar

200 ml Sauternes wine

300 ml carrot juice

2 tsp finely chopped sage

1 tbsp toasted pine nuts

Roll out the shortcrust pastry and line 4 x 8 cm tartlet tins. Prick the pastry with a fork and chill for 30 minutes. Preheat the oven to 200°C/gas mark 6. Line the pastry with greaseproof paper and baking beans. Bake the tartlet cases blind for 15 minutes, then remove the paper and beans and bake for another 5 minutes. Reduce the oven temperature to 170°C/ gas mark 3.

Slice the carrots with the slicing drum on the rotor vegetable slicer/ shredder on speed 4. Cook the carrots in the carrot juice and water until the carrots are tender and the liquid has all but evaporated. Purée the carrots in the blender, then add the cream and process again until smooth. Mix in the egg and egg yolk, then season to taste. Pour into the tartlet cases and bake for 15 to 20 minutes. Leave to cool for 10 minutes.

Sauté the girolle mushrooms in 50 ml olive oil. Season and keep warm. Make the vinaigrette. Sauté the diced carrots in 50 ml olive oil until softened, then add the sprig of sage, salt and balsamic vinegar. Reduce by half. Add the Sauternes and reduce by two thirds. Add the carrot juice and reduce by two thirds, as well. Stir in the remaining olive oil, the chopped sage and the pine nuts. Season to taste. Serve the carrot tartlets with the sautéed girolle mushrooms and lukewarm vinaigrette.

# 4.

# MAIN COURSES

# LINGUINE ALLA BOTTARGA

**Serves 4**

**Prep: 20 minutes**

**Dry: 30 minutes**

**Cook: 10 minutes**

**1 quantity Pasta dough**

**40 g bottarga**

**120 ml extra virgin olive oil**

**¼ tsp chilli flakes**

**1 tbsp finely chopped flat-leaf parsley**

**2 tbsp lemon juice**

Attach the spaghetti cutter to the mixer and feed the sheets of pasta through on speed 2, catching the strands of pasta in one hand as they come through. Lay the strands in a single layer on a clean tea towel or hang them over a pasta drying rack. Leave to dry for 30 minutes, then use. Or coat lightly in flour and store in an airtight tin.

Grate the bottarga with the fine shredding drum on the rotor vegetable slicer/ shredder on speed 4 and set aside. Heat the olive oil in a heavy-based pan and gently cook the chilli flakes for 1 minute. Remove from the heat and stir in the parsley.

Cook the linguine in plenty of lightly salted boiling water until al dente. Drain, then add to the pan with the flavoured oil. Add the bottarga, and toss until each strand of pasta is coated with oil and bottarga. Stir in the lemon juice and serve immediately.

## TAGLIATELLE WITH CRAB, LEMON AND FRESH HERBS

**Serves 4**

**Prep: 20 minutes**

**Dry: 30 minutes**

**Cook: 15 minutes**

**1 quantity Pasta dough, flavoured with**

**1 tbsp Espelette pepper**

**125 ml olive oil**

**2 garlic cloves**

**2 mild red chillies**

**300 g fresh white crabmeat**

**2 tbsp finely chopped flat-leaf parsley**

**1 tbsp finely chopped tarragon**

**2 tbsp finely chopped dill**

**1 tbsp finely chopped mint**

**1 tbsp finely chopped chives**

**juice of 1-2 lemons**

**salt and freshly ground black pepper**

Attach the tagliatelle cutter to the mixer and feed the sheets of pasta through on speed 2, catching the strands of pasta in one hand as they come through.

Lay the strands in a single layer on a clean tea towel or hang them over a pasta drying rack. Leave to dry, then use. Or coat lightly in flour and store in an airtight tin.

Heat the olive oil in a heavy-based pan. Slice the garlic and chillies with the slicing drum on the rotor vegetable slicer/shredder on speed 4. Sauté the garlic and chillies for 5 minutes until lightly coloured.

Remove from the heat and leave to infuse for 10 minutes. Stir in the crabmeat, herbs and lemon juice to taste. Season.

Cook the tagliatelle in plenty of salted boiling water until al dente. Drain but leave a tablespoon or two of the cooking water in the pan. Add the tagliatelle and crab sauce, then toss until every strand of pasta is coated with the sauce. Serve immediately.

## FOUR MUSHROOM RAVIOLI WITH SAGE BUTTER

**Serves 4**

**Prep: 45 minutes**

**Rest: 30 minutes**

**Cook: 15 minutes**

**100 g shallots**

**1 bunch of spring onions**

**2 garlic cloves**

**50 g butter**

**a handful of flat-leaf parsley**

**200 g chestnut mushrooms**

**200 g shiitake mushrooms 200 g oyster mushrooms**

**100 ml Noilly Prat**

**1 quantity Pasta dough, flavoured with 10 g finely ground dried porcini and rolled out to setting 8 thickness**

**125 g butter**

**1 bunch of sage**

**freshly grated Parmesan cheese, to serve**

**salt and freshly ground black pepper**

Finely chop the shallots, spring onions and garlic with the coarse shredding drum on the rotor vegetable slicer/shredder on speed 4. Melt the butter in a large frying pan and sauté the vegetables for 5 minutes until softened. Chop the mushrooms with the coarse shredding drum. Add to the pan and cook on a high heat until the mushrooms have coloured and their juices have evaporated. Deglaze with the Noilly Prat, cook down and leave to cool. Finely chop the parsley and mix into the mushroom mixture. Season to taste.

Lay the pasta sheets on a floured surface. Stamp out rounds with a 5 cm plain or fluted cutter. Fill a piping bag fitted with a large plain nozzle with the mushroom mixture and pipe teaspoonfuls of filling on half of the pasta rounds. Or use a teaspoon to place the filling on the pasta rounds. Moisten the edges of the remaining pasta rounds and place on top to encase the filling. Pinch the edges of the pasta firmly together but make sure that no air is trapped within the ravioli, otherwise they will burst open during cooking. Sprinkle the ravioli with flour and leave to rest for 30 minutes.

Cook the ravioli in plenty of salted simmering water for 3 to 4 minutes or until al dente. Meanwhile, melt the butter over a medium heat. Pick the leaves from the sage. Add to the butter and cook until crispy. Drain the ravioli well and carefully coat in the sage butter. Serve at once with freshly grated Parmesan cheese.

# MUSTARD TORTELLINI WITH SMOKED SALMON AND LEEK CREAM

**Serves 4-6**

**Prep: 40 minutes**

**Rest: 30 minutes**

**Cook: 10 minutes**

**100 g smoked salmon**

**150 g gravlax (Swedish dill-cured salmon)**

**200 g cream cheese**

**¼ tsp cayenne pepper**

**grated zest of 1 lemon**

**2 tbsp finely chopped dill**

**1 egg**

**1 quantity Pasta dough, flavoured with 1 tbsp wholegrain mustard and rolled out to setting 7 thickness**

**2 leeks**

**50 g butter**

**100 ml Noilly Prat**

**250 ml crème fraîche**

**1 tbsp wholegrain mustard**

## salt and freshly ground black pepper

With tortellini, just like ravioli, the choice of ingredients is up to you. Just make sure that you choose a really flavoursome filling.

Grind the smoked salmon and gravlax into the mixer bowl with the fine grinding plate on the food grinder on speed 4. Change to the flat beater and mix in the cream cheese, cayenne pepper, lemon zest, half the dill and the egg on speed 2. Season to taste.

Lay the pasta sheets on a floured surface and cut out 6-7 cm squares. Put the salmon filling in a piping bag fitted with a large plain nozzle and pipe teaspoonfuls of filling in the centre of the pasta squares. Or use a teaspoon to place the filling on the pasta squares. Moisten the edges of the pasta with a little water and fold over to encase the filling and create triangles. Then curve the edges around the filling and pinch them together to form a little mitres (hats). Dust the tortellini with flour and leave to dry for 30 minutes.

Meanwhile, slice the leeks with the slicing drum on the rotor vegetable slicer/shredder on speed 4. Melt the butter in a saucepan and sauté the leeks for 5 minutes until softened. Add the Noilly Prat and reduce by half. Add the crème fraîche and reduce until thickened. Remove from the heat, stir in the mustard and remaining dill. Season to taste and keep warm.

Cook the tortellini in plenty of salted simmering water for 3 to 4 minutes until al dente. Drain the tortellini, then return them to the pan. Gently fold in the leek cream and serve at once. Garnish with a sprinkling of cayenne pepper.

## PUMPKIN GNOCCHI WITH ROCKET AND GOAT'S CHEESE PESTO

**Serves 4**

**Prep: 45 minutes**

**Chill: overnight**

**Cook: 10 minutes**

**750 g pumpkin**

**2 tbsp butter**

**175 g flour**

**1 egg yolk**

**1 quantity Rocket and goat's cheese pesto**

**melted butter and Parmesan shavings, to serve**

**salt and freshly ground black pepper**

Peel the pumpkin and cut into chunks. Roast the pumpkin for 30 minutes in the oven at 200°C/gas mark 6, then purée with the fruit and vegetable strainer on speed 4. Place the pumpkin purée in a pan with the butter and cook on a high heat until the pumpkin has dried out. Return to the mixer bowl and add the flour and egg yolk. Season to taste, then mix with the flat beater on speed 2 to a soft dough. Chill overnight.

The next day, turn out the dough onto a heavily floured surface and roll out into long sausages about 1.5 cm in diameter. Cut into 2 cm pieces and lightly press down with a fork on the gnocchi to make the characteristic pattern.

Bring a large pan of salted water to the boil and cook the gnocchi in batches. Drain the gnocchi well and keep them warm while you cook the remainder. When all the gnocchi are cooked, toss them first in melted butter and then fold in the pesto. Serve at once with Parmesan shavings.

# QUICHE LORRAINE

**Serves 4-6**

**Prep: 10 minutes**

**Cook: 50 minutes**

**1 quantity Shortcrust pastry**

**25g butter**

**250g diced smoked bacon**

**4 eggs**

**150 ml milk**

**150 ml double cream**

**a pinch of freshly grated nutmeg**

**2 tbsp snipped chives**

**salt and freshly ground black pepper**

Preheat the oven to 200°C/gas mark 6. Roll out the pastry into a circle and line a greased 23 cm tart tin. Prick the base all over with a fork and line with greaseproof paper. Fill with baking beans and bake blind for 15 minutes.

Afterwards, remove the paper and baking beans. Lower the oven temperature to 180°C/gas mark 4.

Make the filling. Melt the butter in a large frying pan and sauté the bacon until golden brown. Drain the bacon on kitchen paper, then scatter over the pastry. Mix the eggs, milk, cream, nutmeg and chives in the mixer bowl with the wire whisk on speed 4. Pour over the bacon and bake for 35 to 45 minutes until the top is golden brown and the filling has just set. Serve warm or at room temperature with green salad leaves.

## TORTA PASQUALINA

**Serves 10**

**Prep: 1 hour**

**Rest: 1 hour**

**Cook: 1 hour 20 minutes**

**500 g flour**

**1 tsp salt**

**2 tbsp olive oil approx**

**300 ml water**

**Spinach filling:**

**1 kg fresh spinach**

**2 tbsp olive oil**

**1 small red onion**

**1 bunch of marjoram**

**75 g toasted pine nuts**

**1 large stale ciabatta roll**

**100 ml milk**

**75 g Parmesan cheese**

**8 eggs**

**500 g ricotta**

**6 tbsp olive oil**

**50 g butter**

**salt and freshly ground black pepper**

This savoury Italian Easter tart from Liguria features eggs, a potent Easter symbol. For a quicker and easier version, you could replace the pastry with puff pastry or even filo pastry.

Sieve the flour and salt into the mixer bowl. Mix with the dough hook on speed 1, then slowly add the olive oil and water until you obtain a smooth dough. Continue kneading on speed 2 for 3 minutes until the dough is very smooth and elastic. Divide into 12 pieces, shape into balls and place on a floured tea towel. Cover with a damp tea towel and leave for 1 hour.

Prepare the filling. Remove the stalks from the spinach and wash thoroughly. Cook the spinach in 1 tablespoon of olive oil until just wilted. Drain thoroughly and squeeze out any excess liquid, then chop. Finely chop the onion and sauté in 1 tablespoon of olive oil. Mix into the spinach with the finely chopped marjoram and the pine nuts. Cut the crusts off the bread and tear the bread into pieces. Pour over the milk and leave to soak.

Grate the Parmesan cheese into a small bowl with the fine shredding drum on the rotor vegetable slicer/shredder on speed 4. Break 2 eggs into the mixer bowl and mix with the flat beater on speed 4. Stir in 2 tablespoons of grated Parmesan. Squeeze out any excess moisture from the bread and mix into the eggs with the ricotta. Add the spinach on speed 4.

Preheat the oven to 200°C/gas mark 6. Grease a deep 20-22 cm springform tin. Roll out one of the dough balls on a lightly floured surface, stretching the dough by hand in all directions like strudel pastry. Arrange in the base of the tin, so that the pastry hangs over the sides.

Brush with olive oil. Repeat with the next 5 balls of pastry, brushing each layer with olive oil.

Spoon the filling into the tin, smooth the top and brush with olive oil. Make 6 indentations in the filling. Place a little butter in each and crack an egg on top; take care not to break the yolks. Season, then sprinkle the remaining Parmesan on top.


Roll out the remaining balls of pastry and arrange on top, brushing each layer with olive oil. Place the remaining butter around the edges, then fold over the overhanging pastry. Brush with olive oil and carefully pierce two or three times, so the steam can escape. Bake for 1 hour 15 minutes until golden brown. Serve warm or at room temperature.

# PIZZA RUSTICA

**Makes 2 pizzas**

**Prep: 20 minutes**

**Rest: 1 hour**

**Cook: 20 minutes**

**15 g fresh yeast**

**250 ml lukewarm water**

**a pinch of sugar**

**1 tbsp olive oil**

**400 g white bread flour**

**½ tbsp salt**

**Topping:**

**1 small courgette**

**1 yellow pepper**

**125 g mushrooms**

**3 tbsp olive oil**

**1 jar artichokes in olive oil**

**½ quantity Tomato sauce, or 400 g tinned chopped tomatoes**

**1 tsp dried oregano**

**a handful of black olives**

**250 g smoked mozzarella**

**small basil leaves, to garnish**

**salt and freshly ground black pepper**

Crumble the yeast into a measuring jug and add the water, sugar and olive oil. Stir until the yeast has dissolved and leave for 10 minutes until the mixture starts to foam. Place the flour and salt in the mixer bowl. Mix together with the flat beater on speed 2.

Change to the dough hook and gradually add the yeast mixture on speed 2. Knead for 1 minute until the dough forms a ball. Cover with a damp tea towel and leave to rise for 1 hour, or until doubled in volume.

Prepare the topping. Dice the courgette, slice the pepper thinly and chop the mushrooms. Heat the olive oil in a large frying pan and sauté the vegetables on a high heat until al dente. Drain the artichokes. Add to the vegetables and season to taste. Leave to cool.

Preheat the oven to 200°C/gas mark 6. Knock back the dough, then knead briefly on speed 2. Divide the dough into two pieces and roll each one out into a thin circle. Grease two pizza pans and place the dough in the pans. Make a slightly thicker rim around the edge of the dough.

Spread the tomato sauce or chopped tomatoes over the dough, then top with the vegetables. Sprinkle over the oregano and olives, then finish with the diced mozzarella. Bake for 15 to 20 minutes until the dough is cooked and the cheese is golden brown and bubbling. Garnish with basil leaves and serve at once.

# SPINACH, RICOTTA AND WALNUT CANNELLONI

**Serves 4-6**

**Prep: 30 minutes**

**Cook: 35 minutes**

**600 g fresh spinach**

**75 g shallots**

**100 g walnuts**

**40 g butter**

**2 tbsp brandy**

**400 g ricotta**

**2 eggs**

**75 ml double cream**

**a pinch of freshly grated nutmeg**

**100 g Pecorino Romano cheese**

**75 g Parmesan cheese**

**½ quantity Pasta dough, rolled out to setting 7 thickness**

**1 quantity Tomato sauce**

**salt and freshly ground black pepper**

Spinach and ricotta is a classic combo for vegetarian pasta dishes. Here, it is livened up with the addition of crunchy walnuts and salty cheese.

Wash the spinach and discard any tough stalks. Blanch the spinach briefly in salted boiling water, then plunge immediately into iced water and squeeze out all the excess liquid. Chop the spinach finely.

Chop the shallots and walnuts with the medium shredding drum on the rotor vegetable slicer/shredder on speed 4. Melt the butter in a large frying pan and sauté the shallots and walnuts for 5 minutes until golden brown. Deglaze with the brandy and stir in the spinach. Place the contents of the pan in the mixer bowl. Mix the ricotta, eggs, cream and nutmeg into the spinach mixture with the flat beater on speed 2. Season to taste. Grate the Parmesan and Pecorino with the fine shredding drum on the rotor vegetable slicer/shredder on speed 4. Fold two thirds of the cheese into the spinach mixture.

Preheat the oven to 200°C/gas mark 6. Lay the pasta sheets on a floured surface and cut into rectangles of 15 cm x 7 cm; you need approximately 12 pasta sheets. Cook the pasta sheets in batches for 30 seconds at a time in plenty of salted boiling water. Afterwards, refresh the pasta in cold water with a little olive oil and drain on clean tea towels. Cover to prevent the pasta drying out.

Place about 2 tablespoons of spinach filling at one long end of a pasta sheet. Moisten the edges and roll up to enclose the filling. Place the roll, seam-side down, in a greased ovenproof dish. Repeat with the remaining filling and pasta sheets. Pour over the tomato sauce, then sprinkle with the remaining cheese. Bake for 25 minutes until golden brown and bubbling.

# LOBSTER AND ASPARAGUS CANNELLONI

**Serves 4**

**Prep: 30 minutes**

**Cook: 30 minutes**


**1 kg asparagus, preferably white but green will do**

**12 spring onions**

**25 g butter**

**100 ml champagne**

**2 x 600 g lobsters, freshly boiled**

**3 tbsp crème fraîche**

**3 tbsp vegetable oil**

**100 g celery**

**50 g carrot**

**800 ml lobster stock**

**400 ml double cream**

**50 g melted butter**

**½ quantity Pasta dough, flavoured with 3 capsules powdered saffron and rolled out to setting 7 thickness**

**2 tbsp finely chopped tarragon**

## sea salt and freshly ground black pepper

Peel the asparagus, then slice into small rounds with the slicing drum on the rotor vegetable slicer/shredder on speed 4. Chop the spring onions. Sauté the asparagus and spring onions in the butter until softened.

Add the champagne and cook on a high heat for 5 minutes until the asparagus is tender and the liquid has reduced down. Leave to cool. Crack the lobster shells and remove the meat. Cut the lobster meat into 1 cm dice and mix with the asparagus. Fold in the crème fraîche and season to taste.

Heat the oil in a large pan and sauté the lobster shells on a high heat. Chop the celery and carrot with the coarse shredding drum on the rotor vegetable slicer/shredder on speed 4, then add to the pan and cook for 5 minutes until softened. Add the lobster stock and cream and reduce by half. Sieve the sauce.

Preheat the oven to 200°C/gas mark 6. Lay the pasta sheets on a floured surface and cut into rectangles of 12 cm x 6 cm. Cook the pasta sheets in batches for 30 seconds at a time in plenty of salted boiling water.

Afterwards, refresh the pasta in cold water with a little olive oil and drain on clean tea towels. Cover to prevent the pasta drying out.

Place about 2 tablespoons of lobster filling at one end of a pasta sheet. Moisten the edges and roll up to enclose the filling. Place the cannelloni, seam-side down, in a greased ovenproof dish. Repeat with the remaining filling and pasta sheets. Pour over the melted butter and bake for 20 minutes, basting regularly with the butter.

Meanwhile, gently heat the lobster sauce and stir in the tarragon. Serve the cannelloni as they are or slice them into rounds and serve them cut- side up. Spoon the sauce over the cannelloni.

## LASAGNE ALLA BOLOGNESE

### Serves 4-6

### Prep: 30 minutes

### Cook: 35 minutes

**½ quantity Pasta dough, rolled out to setting 6 thickness**

**50 g Parmesan cheese**

**1 quantity Béchamel sauce**

**2 tbsp finely chopped rosemary**

**1 quantity Ragù**

**salt and freshly ground black pepper**

Preheat the oven to 200°C/gas mark 6. Lay the pasta sheets on a floured surface and cut into rectangles of 15 cm x 7 cm; you need approximately 12 pasta sheets. Cook the pasta sheets in batches for 30 seconds at a time in plenty of salted boiling water. Afterwards, refresh the pasta in cold water with a little olive oil and drain on clean tea towels. Cover to prevent the pasta drying out.

Grate the Parmesan cheese with the fine shredding drum on the rotor vegetable slicer/shredder on speed 4. Stir half the Parmesan cheese into the béchamel sauce. Stir the rosemary into the ragù.

Place a layer of pasta sheets in the base of a greased ovenproof dish. Spoon over half the ragù, a third of the béchamel sauce and cover with a layer of pasta sheets. Repeat these layers and finish with the remaining béchamel sauce and Parmesan cheese. Bake for 20 to 25 minutes until golden brown and bubbling.

## OPEN LASAGNE WITH ROASTED VEGETABLES AND HERBED RICOTTA

**Serves 4-6**

**Prep: 30 minutes**

**Cook: 20 minutes**

**1 quantity Pasta dough, prepared with 4 tbsp Pesto instead of 1 egg and rolled out to setting 7 thickness**

**2 small courgettes**

**1 red pepper**

**1 yellow pepper**

**1 fennel bulb**

**250 g cherry tomatoes**

**4 tbsp olive oil**

**25 g Parmesan cheese**

**1 bunch of marjoram or basil**

**2 tbsp extra virgin olive oil**

**250 g ricotta**

**extra virgin olive oil, for drizzling**

**salt and freshly ground black pepper**

Replacing the cooked tomato sauce with cherry tomatoes and not cooking the lasagne in the oven makes for a lighter dish, ideal for summer.

Preheat the oven to 220°C/gas mark 7. Lay the pasta sheets on a floured surface and cut into rectangles of 12 cm x 10 cm. Cook the pasta sheets in batches for 30 seconds at a time in plenty of salted boiling water.

Afterwards, refresh the pasta in cold water with a little olive oil and drain on clean tea towels. Cover to prevent the pasta drying out and keep warm.

Slice the courgettes, peppers and fennel with the slicing drum on the rotor vegetable slicer/shredder on speed 4. Place the vegetables in a roasting tray and pour over the olive oil. Roast the vegetables for

15 minutes until they begin to soften. Add the cherry tomatoes and roast for another 5 minutes. Remove from the oven.

Grate the Parmesan cheese with the fine shredding drum on the rotor vegetable slicer/shredder on speed

4. Roughly chop the marjoram or basil. Mix the Parmesan, marjoram or basil and extra virgin olive oil into the ricotta cheese. Season to taste.

Place a sheet of pasta on each plate and top with roasted vegetables. Dot with some of the ricotta mixture and cover with another pasta sheet. Repeat these layers and drizzle some extra virgin olive oil over the top. Serve at once.

## CHAMPAGNE-POACHED SALMON WITH PICKLED CUCUMBER

**Serves 6-8**

**Prep: 50 minutes**

**Rest: 2 hours**

**Cook: 1 hour 5 minutes**

**1 x 2 kg whole salmon, scaled and gutted**

**1 onion**

**1 lemon**

**2 celery stalks**

**½ fennel bulb**

**a handful of parsley stalks**

**2 bay leaves**

**400 ml champagne**

**750 ml water**

**Pickled cucumber:**

**3 large shallots**

**3 large cucumbers**

**50 g fine sea salt**

**500 g sugar**

**500 ml white wine vinegar**

**1 tbsp yellow mustard seeds**

**1 tsp coriander seeds**

**10 black peppercorns**

**2 bay leaves**

First make the pickled cucumber. Slice the shallots and cucumbers with the slicing drum on the rotor vegetable slicer/shredder on speed 4. Layer the vegetables in a colander and sprinkle salt on each layer. Place a plate on top and weigh down with a few tins. Leave for 2 hours, then squeeze out as much liquid as possible.

In a large saucepan bring the sugar and white wine vinegar to the boil with the spices and bay leaves. Stir until the sugar has dissolved. Add the drained vegetables and simmer, uncovered, for 1 minute. Remove from the heat, drain the vegetables and reserve the liquid. Spoon the vegetables into hot sterilized jars. Bring the liquid back to the boil and reduce for 15 minutes, then pour onto the vegetables. Leave to cool completely, then seal the jars.

Lightly rinse the salmon and pat dry with kitchen paper. Leave the skin on: this will help keep the fish intact as it is cooking. Slice the onion, lemon, celery and fennel with the slicing drum on the rotor vegetable slicer/shredder on speed 4. Place in a fish kettle or large saucepan with the parsley stalks, bay leaves and champagne. Pour in the water and bring to the boil. Reduce the heat and simmer for 10 minutes.

Place the salmon on the vegetables, bring back to the boil and simmer the salmon for 10 minutes per 450 g weight. Afterwards, remove the fish kettle or pan from the heat but leave the salmon to cool in its own liquor.

When cold, remove the salmon from the fish kettle or pan and carefully peel off the skin, discarding the eyes and gills in the process. Snip the backbone at the head and tail with scissors, then carefully ease the flesh away from the bone. Serve with the pickled cucumber.

## CHERMOULA TUNA BURGERS WITH PRESERVED LEMON COUSCOUS

**Serves 4**

**Prep: 10 minutes**

**Marinate: 15 minutes**

**Chill: 1 hour**

**Cook: 5 minutes**


**1 small red onion**

**2 garlic cloves**

**1 tsp ground cumin**

**1 tsp paprika**

**2 capsules powdered saffron**

**¼ tsp cayenne pepper**

**4 tbsp lemon juice**

**6 tbsp olive oil**

**1 bunch of coriander**

**½ bunch of flat-leaf parsley**

**600–700 g fresh tuna**

**salt and freshly ground black pepper**


**Preserved lemon couscous:**

**200 g couscous**

**75 g raisins**

**250 ml hot chicken stock**

**75 g shelled pistachio nuts**

**1 preserved lemon**

**4 tbsp extra virgin olive oil**

**a squeeze of lemon juice**

**a handful of coriander**


Chermoula is a Moroccan herb paste which can be used as a marinade, a dressing for salads or couscous, or to flavour these tuna burgers. Be careful not to overcook the burgers or the tuna will dry out and the chermoula will lose its fragrance.

First make the chermoula. Grate the onion and garlic into the mixer bowl with the medium shredding drum on the rotor vegetable slicer/shredder on speed 4. Stir in the spices, lemon juice and 3 tablespoons of olive oil. Leave for 15 minutes.

Make the preserved lemon couscous. Put the couscous and raisins in a bowl and pour over the hot stock. Cover and leave for 15 minutes, or until the couscous and raisins have absorbed the stock. Fluff up the couscous with a fork. Roughly chop the pistachios and preserved lemon peel (discard the flesh). Stir into the couscous with the extra virgin olive oil and season with lemon juice. Finely chop the coriander and fold into the couscous. Season to taste.

Grind the tuna into the mixer bowl with the coarse grinding plate on the food grinder on speed 4. Add three quarters of the chermoula. Change to the flat beater and mix on speed 2 until the tuna and chermoula are well- blended. Season to taste. Finely chop the coriander and parsley. Mix into the tuna on speed 1. Shape the tuna mixture into 4 burgers, cover and chill for 1 hour. Bring back to room temperature before cooking.

Heat the remaining olive oil in a non-stick pan and fry the tuna burgers on a high heat for 1 to 2 minutes on each side; they should be golden brown on the outside but rare to medium rare in the middle. Serve at once with the couscous and the remaining chermoula.

## FISH 'N' CHIPS WITH MINTED PEA PURÉE

**Serves 4**

**Prep: 10 minutes**

**Cook: 35 minutes**

**100 g flour**

**100 g cornflour**

**½ tsp salt**

**200-300 ml ice-cold lager**

**600 g white fish fillets, eg sole, brill or plaice
sunflower oil, for deep frying**

**1 quantity of Frites, to serve**

**sea salt flakes and white balsamic vinegar, for sprinkling**

**Pea purée:**

**2 shallots**

**25 g butter**

**300 g frozen peas**

**½ bunch of mint**

**a good pinch of sugar**

**150 ml dry white wine**

**150 ml double cream**

**salt and freshly ground black pepper**

Sieve the flour, cornflour and salt into the mixer bowl. Slowly whisk in enough lager with the wire whisk on speed 4 to make a smooth batter. Cover and chill.

Finely chop the shallots with the medium shredding drum on the rotor vegetable slicer/shredder on speed 4 and sauté them in the butter until softened. Stir in the peas, half the mint leaves and the sugar. Season to taste, then add the white wine and cover the pan. Gently cook for 30 minutes until the peas are tender.

Uncover the pan, pour in the cream and cook over a medium to high heat until most of the cream has reduced down. Remove from the heat and pour into the blender. Process on purée speed until smooth. Add the remaining mint and blend again until flecks of mint are still visible in the pea purée. Keep warm.

Heat the sunflower oil in a deep fryer to 180°C. Slice the fish into 8 pieces and dip each into the batter until the fish is well-coated. Fry in two batches for 2 to 3 minutes until crispy and golden brown. Remove with a slotted spoon and drain on kitchen paper before sprinkling with sea salt and white balsamic vinegar.

Serve with the pea purée and frites.

## TANDOORI SALMON WITH FRESH COCONUT, CUCUMBER AND MINT SALAD

**Serves 4**

**Prep: 10 minutes**

**Marinate: 1 hour**

**Cook: 10 minutes**

**1 ½ tbsp coriander seeds**

**1 tbsp cumin seeds**

**1 tbsp paprika**

1 red bird's-eye chilli

2 large garlic cloves

5 cm fresh ginger

2 tbsp lemon juice

1 tsp garam masala (Indian spice mix)

¼ tsp turmeric

200 ml natural yoghurt

4 x 175 g skinless salmon fillets

1 tbsp groundnut oil

lime wedges, to serve

salt and freshly ground black pepper


Salad:

75 g fresh coconut

¼ cucumber

1 mild green chilli

½ tsp salt

2 tbsp finely chopped mint

2 tbsp raisins

2 tbsp sunflower oil

1 tsp black mustard seeds

Toast the coriander and cumin seeds in a dry frying pan until they start to release their aroma. Remove from the pan and cool, then grind to a fine powder with a pestle and mortar or in a coffee mill.

Mix with the next 8 ingredients in the blender on stir speed, then season. Rinse the salmon and pat dry with kitchen paper. Place in a shallow dish and pour over the marinade, massaging it into the fish with your hands. Cover and marinate for at least 1 hour. Baste the salmon occasionally with the marinade.

Make the salad. Grate the coconut into the mixer bowl with the fine shredding drum on the rotor vegetable slicer/shredder on speed 4. Deseed the cucumber, then grate with the coarse shredding drum. Deseed and finely chop the chilli. Add to the mixer bowl with the salt, mint and raisins. Mix with the flat beater on speed 1 until combined. Heat the oil in a small saucepan and fry the mustard seeds until they start to pop. Pour into the salad and mix again on speed 2.

Preheat the oven to 240°C/gas mark 8. Bake the salmon for 8 to 10 minutes or until just cooked. Serve with the salad, lime wedges and boiled rice.

# SEARED SEA BASS WITH BLOOD ORANGE HOLLANDAISE

**Serves 4-6**

**Prep: 5 minutes**

**Cook: 20 minutes**

**1 shallot**

**100 ml freshly squeezed blood orange juice**

**1 tsp honey (optional)**

**4 egg yolks**

**250 g warm melted butter, preferably clarified**

**a squeeze of lemon juice**

**4-6 x 150 g sea bass fillets with the skin on**

**1 tbsp olive oil**

**1 tbsp butter**

**salt and freshly ground white pepper**

Slice the shallot with the slicing drum on the rotor vegetable slicer/shredder on speed 4. Place in a small saucepan with the blood orange juice and honey if using, bring to the boil and reduce by half. Strain into the mixer bowl.

Add the egg yolks and whisk with the wire whisk on speed 8 until frothy. Add the melted butter to the mixer bowl as the motor is running: drop by drop at first, then in a slow drizzle as the hollandaise begins to thicken. Season with salt and pepper, and a squeeze of lemon juice. Cover to keep warm.

Season the sea bass fillets on the flesh side only. Heat the olive oil and butter in a non-stick frying pan, then fry the sea bass skin-side down on a high heat until just cooked. Turn over and cook for 15 seconds on the flesh side. Place the sea bass skin-side up on warm plates, season to taste and spoon the hollandaise over and around the fish. Serve with steamed green asparagus or grilled fennel.

# FINNISH MEATBALLS WITH HORSERADISH CREAM AND CRANBERRY COMPOTE

**Serves 6**

**Prep: 30 minutes**

**Soak: 30 minutes**

**Chill: 20 minutes**

**Cook: 20 minutes**

3 slices of stale white bread

150 ml milk

500 g lean beef

500 g pork

1 red onion

1 large egg

1 ½ tsp ground allspice

2 tbsp sunflower oil

60 g butter

1-2 tbsp flour

400 ml chicken stock

200 g soured cream

**grated zest of 1 lemon**

**1-2 tbsp horseradish cream**

**salt and freshly ground black pepper**

**Cranberry compote:**

**500 g cranberries (or lingonberries)**

**1 finely chopped red onion**

**grated zest and juice of 1 lemon**

**125 ml water**

**200 g sugar**

First make the cranberry compote. Put the cranberries in a pan with the onion, lemon zest and juice, and the water. Bring to the boil, then simmer for 15 minutes until the compote has thickened. Add the sugar and stir until dissolved, then remove from the heat and leave to cool to room temperature.

Remove the crusts from the bread and soak the bread in the milk for 30 minutes until the bread has absorbed all the milk. Mince the beef, pork and onion into the mixer bowl with the fine grinding plate on the food grinder on speed 4. Add the egg, allspice and soaked bread. Change to the flat beater, then mix on speed 4 until well-blended. Season to taste, then chill for 20 minutes.

Shape the meat mixture into balls the size of walnuts. Heat the oil and 40 g butter in a large frying pan and fry the meatballs in batches until golden brown and cooked through. Keep warm. When all the meatballs are cooked, pour away most of the fat. Add the remaining butter to the frying pan and stir in the flour. Cook for 2 minutes until the roux is golden, then stir in the stock until you obtain a smooth sauce. Add the soured cream, lemon zest and horseradish sauce to taste and mix well. Return the meatballs to the pan and coat them in the sauce. Serve with the cranberry compote and some boiled rice.

# POLPETTE WITH MASCARPONE TOMATO SAUCE

**Serves 4**

**Prep: 20 minutes**

**Chill: 20 minutes**

**Cook: 30 minutes**

**50 g stale white bread**

**4 tbsp milk**

**400 g lean beef**

**50 g Parmesan cheese**

**½ small red onion**

**1 garlic clove**

**3 tbsp olive oil**

**1 tsp fennel seeds**

**grated zest of 1 lemon**

**1 egg**

**½ quantity of Tomato sauce, flavoured with 2 tbsp finely chopped basil**

**125 g mascarpone**

**salt and freshly ground black pepper**

Remove the crusts from the bread and sprinkle the milk over the bread. Soak for 10 minutes until the bread has absorbed all the milk. Place in the mixer bowl. Mince the beef into the mixer bowl with the fine grinding plate on the food grinder on speed 4.

Grate the Parmesan cheese with the fine shredding drum on the rotor vegetable slicer/shredder on speed

4. Chop the onion and garlic with the medium shredding drum. Heat 1 tablespoon of olive oil in a frying pan and sauté the onion and garlic for 5 minutes until softened. Leave to cool for 5 minutes, then add to the mixer bowl with the Parmesan cheese, fennel seeds, lemon zest and egg. Change to the flat beater and mix on speed 4 until well-blended. Season to taste and chill for 20 minutes.

Shape the mixture into balls the size of cherry tomatoes. Heat the remaining olive oil in a large frying pan and fry the meatballs in batches until golden brown and cooked through. Gently heat up the tomato sauce and stir in the mascarpone. Mix into the meatballs and serve with spaghetti or tagliatelle.

## CHICKEN AND SEAFOOD LAKSA

**Serves 4**

**Prep: 15 minutes**

**Cook: 25 minutes**

**150 g rice noodles**

**1 red pepper**

1 yellow pepper

125 g mangetout

1 onion

1-2 red bird's-eye chillies

2 lemongrass stalks

2 garlic cloves

2.5 cm peeled ginger

25 g macadamia nuts

2 tbsp groundnut oil

1 tsp turmeric

300 g chicken breast fillets

600 ml coconut milk

400 ml chicken stock

12 raw and peeled medium-sized prawns

3-4 tbsp fish sauce

juice of 1 lime

1 bunch of coriander

Laksa is a Malaysian noodle dish, a cross between a soup and a stew, in which vegetables, seafood and/or meat are served with a spicy broth. The latter is often flavoured and coloured with turmeric.

Put the rice noodles in a bowl, cover with boiling water and leave to soak. Slice the peppers and mangetout with the slicing drum on the rotor vegetable slicer/shredder on speed 4.

Replace with the coarse shredding drum and chop the onion. Set aside, then mix the chillies, chopped lemongrass, garlic, ginger and macadamia nuts to a paste in the blender on stir speed. Heat the oil in a wok and sauté the onion for 5 minutes until softened. Add the spice paste and turmeric and cook for another 2 minutes. Slice the chicken into strips or bitesize pieces. Add to the wok and sauté for 3 minutes more.

Pour in the coconut milk and chicken stock, and bring to the boil. Reduce the heat and simmer for 10 minutes until the chicken is cooked. Add the vegetables and prawns, then simmer for 3 minutes until the vegetables are tender to the bite and the prawns are just cooked.

Add the fish sauce and lime juice. Finely chop the coriander and stir through the laksa. Drain the rice noodles thoroughly, then divide between 4 bowls. Ladle the laksa onto the noodles and serve.

# CHICKEN AND APRICOT TAGINE WITH SWEET AND SOUR CARROT SALAD

**Serves 4**

**Prep: 15 minutes**

**Cook: 1 hour**

**2 tbsp olive oil**

**4 skinless chicken legs**

**1 large onion**

**2 garlic cloves**

**½ tsp ground cinnamon**

**1 tsp ground coriander**

**1 tsp ground cumin**

**a large pinch of saffron threads**

**150 g dried apricots**

**50 g sultanas**

**600 ml chicken stock**

**200 g tinned chopped tomatoes**

**400 g tinned chickpeas**

**1 bunch of coriander**

**50 g toasted flaked almonds**

**salt and freshly ground black pepper**

**Carrot salad:**

**5 tbsp olive oil**

**juice of ½ - 1 lemon**

**1 tbsp orange flower water**

**1 tbsp orange flower honey**

**500 g carrots**

**1 tbsp finely chopped coriander argan oil, to taste**

Heat 1 tablespoon of olive oil in a heavy-based pan and brown the chicken on all sides. Remove from the pan. Chop the onion and garlic with the medium shredding drum on the rotor vegetable slicer/shredder on speed 4. Sauté in 1 tablespoon of olive oil for 5 minutes until softened. Add the cinnamon, coriander and cumin and cook for another minute.

Return the chicken to the pan and add the saffron, half the apricots, the sultanas, stock and tomatoes. Bring to the boil, then reduce the heat and cover the pan. Simmer for 40 minutes until the chicken is cooked.

Meanwhile, make the carrot salad. Place the first 4 ingredients in the mixer bowl and mix with the wire whisk on speed 4 until well-blended. Grate the carrots into the mixer bowl with the coarse shredding drum on the rotor vegetable slicer/shredder on speed 4. Mix with the dressing and add the coriander. Season to taste with a few drops of argan oil.

When the chicken is cooked, stir the drained chickpeas and remaining apricots into the tagine and simmer for another 10 minutes until the apricots have softened. Finely chop the coriander and stir into the tagine. Remove from the heat. Spoon onto plates and sprinkle over the almonds. Serve with the carrot salad.

## FILET AMÉRICAIN WITH FRITES

**Serves 4-6**

**Prep: 15 minutes**

**Chill: 1 hour**

**Cook: 40 minutes**

**800 g lean beef steak**

**4 tbsp olive oil**

140

**1 tbsp Dijon mustard**

**1 tbsp Worcestershire sauce**

**¼ - ½ tsp Tabasco**

**60 g gherkins**

**60 g onion**

**60 g capers**

**4-6 very fresh organic egg yolks**

**2 tbsp finely chopped flat-leaf parsley**

**salt and freshly ground black pepper**


**Frites:**

**1 kg floury potatoes groundnut oil, for deep frying**

**sea salt flakes, for seasoning**

Trim the beef, discarding any fat or gristle, and cut into cubes. Mince the meat into the mixer bowl with the coarse grinding plate on the food grinder on speed 4. Do this twice, so the meat is reasonably finely ground.

Change to the flat beater and mix in the olive oil, mustard, Worcestershire sauce and Tabasco to taste on speed 1 until well-blended. Season to taste, divide into 4 to 6 portions and shape into steaks. Cover lightly and chill for 1 hour. Chop the gherkins, onion and capers separately with the medium shredding drum on the rotor vegetable slicer/shredder on speed 4. Place in separate bowls.

Peel and cut the potatoes into 1 cm slices and then into 5 mm strips. Cover with cold water and leave to soak for 1 hour. Afterwards, drain the potatoes thoroughly and pat dry with a clean tea towel. Heat the oil in a deep fryer to 170°C. Deep fry the potatoes in batches for 3 to 5 minutes until they are tender but still pale in colour. Drain on kitchen paper and leave to cool completely.

Just before serving, reheat the oil to 190°C and deep fry the blanched frites in batches for 2 to 4 minutes until they are golden brown and crispy. Drain on kitchen paper and sprinkle with sea salt. Place the filet américain on plates and make a shallow depression in each steak.

Carefully slide an egg yolk in the middle. Garnish with little heaps of gherkins, onion, capers and parsley. Serve with the hot frites.

# LAMB EN CROÛTE WITH MINT AND PISTACHIO PESTO

**Serves 8**

**Prep: 30 minutes**

**Chill: 15 minutes**

**Cook: 20 minutes**

**4 x 225 g lamb fillets, approximately**

**10 cm long**

**125 g butter**

**½ quantity Mint and pistachio pesto**

**1 quantity Flaky pastry**

**1 egg yolk**

**175 ml lamb stock**

**75 ml Marsala wine**

**salt and freshly ground black pepper**

Preheat the oven to 220°C/gas mark 7. Halve the lamb fillets. Melt 50 g butter in a large frying pan and season the lamb. Sear the lamb fillets for 30 seconds on each side over a high heat. Leave to drain on a wire rack but reserve the lamb juices.

Roll out the flaky pastry and cut into 8 rectangles measuring 20 cm x 10 cm. Lay the rectangles on a lightly floured surface and spread 1 teaspoon of pesto on each. Pat the lamb dry with kitchen paper, then place each lamb fillet on a pastry rectangle. Spread another teaspoon of pesto on the lamb and fold the pastry over the meat. Brush the ends with a little cold water and seal the parcels. Turn the parcels over and score the top. Chill for 15 minutes.

Whisk the egg yolk with 1 tablespoon of water and glaze the lamb parcels. Bake the parcels for 12 to 14 minutes until golden brown. Meanwhile, place the lamb juices in a small saucepan with the lamb stock and Marsala wine. Reduce by half, then whisk in the remaining butter until the sauce has thickened and become glossy. Serve the lamb parcels with the Marsala sauce and some roast tomatoes.

# SPICY SAUSAGES WITH LENTILS IN TOMATO SAUCE

**Makes 8 small sausages**

**Prep: 30 minutes**

**Soak: 30 minutes**

**Cook: 10 minutes**

**1 kg pork belly, rind and bones removed**

**4 tsp pimentón (Spanish smoked paprika)**

**1 tsp chilli flakes**

**1 tbsp fennel seeds**

**120 cm natural or synthetic sausage casing**

**4 tbsp olive oil**

**salt and freshly ground black pepper**

**Lentils:**

**150 g pancetta or smoked bacon**

**2 red onions**

**2 garlic cloves**

**250 g mushrooms**

**200 g Puy or brown lentils**

**½ quantity Tomato sauce, flavoured with a generous pinch of saffron threads**

**4 tbsp finely chopped flat-leaf parsley**

Cut the pork into 5 cm cubes and chill thoroughly. Mince the pork into the mixer bowl with the coarse grinding plate on the food grinder on speed 4, then mince again with the fine grinding plate. Add the spices and mix with the flat beater on speed 4, then season to taste.

Soak the natural casing in cold water for 30 minutes, then rinse thoroughly. Grease the large sausage stuffer tube and slide the casing on tightly. Tie off the end. Feed the pork mixture through on speed 4, twisting and shaping the sausages into small links as you go. Chill.

Chop the pancetta and sauté in a large saucepan until the pancetta is golden brown and has released its fat. Chop the onions and garlic with the medium shredding drum on the rotor vegetable slicer/shredder on speed 4. Add to the pancetta and sauté for 5 minutes until softened.

Chop the mushrooms with the coarse shredding drum. Add to the pan and sauté for another 5 minutes. Add the rinsed lentils, barely cover with boiling water and simmer on a medium heat for 20 minutes until the lentils are tender to the bite. Stir in the tomato sauce and parsley, then heat through and season to taste. Heat the olive oil in a frying pan and fry the sausages until golden brown and cooked through. Serve with the lentils.

## BRATWURST WITH SAUERKRAUT RELISH

**Serves 3-4**

**Prep: 15 minutes**

**Cook: 15 minutes**

**300 g lean pork fillet**

**150 g veal escalopes or lean beef**

**3 shallots**

**25 g fresh white breadcrumbs**

**½ tsp ground ginger**

**½ tsp ground coriander**

**¼ tsp ground caraway**

½ tsp freshly grated nutmeg

a pinch of ground cardamom

1 tbsp finely chopped flat-leaf parsley

60 cm natural or synthetic

sausage casing vegetable oil, for frying honey mustard, to serve

salt and freshly ground white pepper


Sauerkraut relish:

250 g sauerkraut

1 onion

2 green celery stalks

½ red pepper

½ green pepper

½ tsp caraway seeds

100 g sugar

100 ml white wine vinegar


First make the relish. Drain the sauerkraut thoroughly, then place in a glass bowl. Chop the onion, celery and peppers with the coarse shredding drum on the rotor vegetable slicer/shredder on speed 4. Mix into the sauerkraut with the caraway seeds. Place the sugar and vinegar in a small saucepan and heat gently, stirring until the sugar has dissolved. Bring to the boil, immediately remove from the heat and pour over the sauerkraut and vegetables. Mix well, cover and leave to marinate.

Trim the meat and cut into 5 cm cubes. Chill the meat thoroughly, then mince into the mixer bowl: first with the coarse grinding plate on the food grinder on speed 4, then again with the fine grinding plate. Finely chop the shallots and add to the bowl with the breadcrumbs, spices and parsley. Mix with the flat beater on speed 4. Season to taste.

Soak the natural casing in cold water for 30 minutes, then rinse thoroughly. Grease the large sausage stuffer tube and slide the casing on tightly. Tie off the end. Feed the pork mixture through on speed 4, twisting and shaping the sausages as you go. Heat some oil in a frying pan and fry the sausages for 10 minutes, or until golden brown and cooked through. Serve with the relish and honey mustard.

# 5.

# SIDE DISHES

# PLUM, PEAR AND SULTANA CHUTNEY

**Makes 1.2 kg**

**Prep: 15 minutes**

**Cook: 1 hour 30 minutes**

**450 g red plums**

**450 g pears**

**225 g onions**

**50 g fresh ginger**

**225 g sultanas**

**grated zest of ½ orange**

**grated zest of ½ lemon**

**570 ml cider vinegar**

**350 g warm light brown sugar**

**½ tsp ground cinnamon**

**¼ tsp ground ginger**

**¼ tsp ground cloves**

**2 tsp salt**

Halve and stone the plums. Peel and core the pears. Slice the plums, pears and onions with the slicing drum on the rotor vegetable slicer/shredder on speed 4. Peel and grate the ginger with the medium shredding drum. Place the fruit, onions and ginger in a large heavy-based saucepan and add the remaining ingredients. Slowly bring to the boil, stirring until the sugar has dissolved. Turn up the heat, then cook for 1 hour 30 minutes until the chutney has thickened. Stir occasionally. Spoon the chutney into hot sterilized jars and seal.

# GRATIN DAUPHINOIS WITH MORELS

**Serves 6**

**Prep: 15 minutes**

**Infuse: 15 minutes**

**Cook: 1 hour 15 minutes**

**500 ml double cream**

**250 ml milk**

**30 g dried morels**

**1 kg waxy potatoes**

**2 garlic cloves**

**25 g softened butter**

**salt and freshly ground black pepper**

Preheat the oven to 180°C/gas mark 4. Bring the cream and milk to the boil, pour over the morels and leave to infuse for 15 minutes. Afterwards, strain the cream into a jug and finely slice the morels. Peel and slice the potatoes with the slicing drum on the rotor vegetable slicer/shredder on speed 4.

Halve the garlic cloves and rub them all over the base and sides of a large ovenproof dish. Grease with the butter. Layer the potatoes and morels in the dish and season each layer generously. Pour the infused cream over the vegetables. Cover with aluminium foil and bake for 1 hour until the potatoes are tender. Remove the foil, then bake for another 15 minutes until the top is golden brown.

# BUBBLE AND SQUEAK RÖSTI

**Serves 4**

**Prep: 15 minutes**

**Cook: 10 minutes**

**75 g Savoy cabbage**

**50 g Cheddar cheese**

**450 g waxy potatoes**

**25 g dried cranberries**

**2 tsp finely chopped sage**

**2 tbsp flour**

**1 egg**

**25 g melted butter**

**2 tbsp vegetable oil**

**salt and freshly ground black pepper**

Slice the cabbage with the slicing drum on the rotor vegetable slicer/shredder on speed 4. Grate the Cheddar with the fine shredding drum and grate the potatoes with the medium shredding drum. Press as much liquid out of the potatoes as possible.

Place the potatoes, cabbage and cheese in a bowl. Add the cranberries, sage, flour and egg. Season to taste. Mix with two forks, then shape into 8 or 12 patties. Heat the butter and oil in a non-stick frying pan on a medium heat. Fry the rösti in batches until golden brown, turning them halfway through the cooking time.

## BRAISED RED CABBAGE WITH RED WINE AND MIXED SPICE

**Serves 4-6**

**Prep: 10 minutes**

**Cook: 2 hours**

**500 g red cabbage**

**1 red onion**

**75 g butter**

**3 tbsp light brown sugar**

**1 large cooking apple**

**100 g dried cranberries**

**375 ml fruity red wine**

**juice of 1 orange**

**½ tsp mixed spice**

**4 juniper berries**

**1 tbsp balsamic vinegar**

**salt and freshly ground black pepper**

Remove the outer leaves and hard inner core of the cabbage. Slice the cabbage and onion with the slicing drum on the rotor vegetable slicer/shredder on speed 4. Melt the butter in a heavy-based pan and sauté the onion until softened. Add the cabbage and stir well, then sprinkle over the sugar.

Peel and core the apple, then grate with the coarse shredding drum on the rotor vegetable slicer/shredder on speed 4. Add to the pan with the cranberries, wine, orange juice and spices. Sauté until the vegetables have wilted a little, then cover and simmer for 2 hours until the vegetables have cooked down and softened. Stir in the balsamic vinegar, season to taste and serve.

# 6.

# BREADS

# SPELT SOURDOUGH

**Makes 1 large loaf**

**Prep: 15 minutes**

**Rise: 2 hours 10 minutes**

**Ferment: 2-3 days**

**Cook: 40 minutes**

**125 g wholemeal spelt flour**

**250 g white bread flour**

**1 ½ tsp salt**

**1 tsp dried yeast**

**175 ml lukewarm water**

**Starter:**

**250 g wholemeal spelt flour**

**7 g dried yeast**

**300 ml water**

First make the starter. Sieve the spelt flour into the mixer bowl; add what is left in the sieve to the bowl, as well. Stir in the yeast, add the water and mix with the flat beater on speed 2. Pour into a bowl, cover loosely and keep in a warm, draught-free place for 2 to 3 days. Give the starter a stir twice a day. The starter is ready when it produces a pleasantly sour smell.

Sieve the two types of flour and salt into the mixer bowl. Add the yeast. Measure 250 ml of the starter into a jug, then stir in the lukewarm water. Gradually knead this mixture into the flour with the dough hook on speed 1 until you obtain a smooth dough. Continue to knead on speed 2 for 3 minutes until the dough is smooth and elastic. Cover with clingfilm and leave to rise in a warm place for 1 hour, or until doubled in volume.

Knock back the dough, then leave to rest for 10 minutes. Form the dough into a round or oval shape and place on a greased baking sheet. Make a few slashes in the top. Cover with oiled clingfilm and leave to rise for 1 hour, or until doubled in volume.

Preheat the oven to 220°C/gas mark 7 and place a shallow dish with water in the bottom. Lightly dust the dough with flour, Bake for 10 minutes, then lower the oven temperature to 200°C/gas mark 6 and bake for another 30 to 40 minutes until golden brown. The bread should sound hollow when tapped on the base.

## GRANARY, WALNUT AND HONEY BREAD

**Makes 2 loaves**

**Prep: 15 minutes**

**Rise: 2 hours**

**Bake: 40 minutes**

**150 g walnuts**

**1 kg granary flour**

**1 tsp salt**

**7 g dried yeast**

**2 tbsp walnut oil**

**5 tbsp heather honey**

**525 ml lukewarm water**

Toast and roughly chop the walnuts. Mix the flour, salt and yeast in the mixer bowl with the flat beater on speed 2. Mix the walnut oil and honey with the water. Change to the dough hook and gradually mix the water mixture into the flour on speed 2. Knead for 1 minute, then place the dough in a greased bowl. Coat the dough in the oil and leave to rise in a warm place for 1 hour, or until doubled in volume.

Mix the walnuts into the dough with the dough hook on speed 1, then knead the dough for 1 more minute. Add a little flour if the dough seems too sticky. Divide the dough into two portions and place each in a 1 kg loaf tin. Cover with a damp tea towel and leave to rise for 1 hour, or until the dough has risen above the rim of the tins.

Preheat the oven to 220°C/gas mark 7. Bake the loaves for 10 minutes, then lower the oven temperature to 200°C/gas mark 6. Bake for another 30 minutes, or until the loaves sound hollow when tapped on the base. Cool on a wire rack.

# GOAT'S CHEESE, SPRING ONION AND THYME SODA BREAD

**Makes 1 loaf**

**Prep: 10 minutes**

**Cook: 35 minutes**

**250 g flour**

**250 g wholemeal flour**

**1 tsp salt**

**1 tsp bicarbonate of soda**

**1 tbsp finely chopped thyme**

**6 finely chopped spring onions**

**300-350 ml buttermilk**

**100 g mature goat's cheese**

Preheat the oven to 230°C/gas mark 8. Sieve the two types of flour, salt and bicarbonate of soda into the mixer bowl; add what is left in the sieve to the bowl, as well. Stir the thyme and spring onions into the flour. Gradually mix in the buttermilk with the flat beater on speed 2 until you obtain a smooth dough; you may not need all of the buttermilk.

Change to the dough hook, then crumble in the goat's cheese and knead on speed 2 until the dough is soft but not sticky. Shape into a 2.5 cm thick round loaf, carve a deep cross into the top and place on a greased baking sheet. Bake for 15 minutes, then lower the oven temperature to 200°C/gas mark 6.

Bake for another 15 to 20 minutes. The bread should sound hollow when it is tapped on the base. Serve warm or at room temperature but eat the same day it is made.

## CUMIN AND CORIANDER PITTA BREAD

**Makes 12**

**Prep: 15 minutes**

**Rise: 2 hours**

**Cook: 10 minutes**

**15 g fresh yeast**

**300 ml lukewarm water**

**1 tbsp sugar**

**1 ½ tbsp olive oil**

**500 g white bread flour**

**2 tsp salt**

**2 tsp ground cumin**

**1 bunch of coriander**

Crumble the yeast into the mixer bowl and add the water. Stir in the sugar and olive oil. Leave for 10 minutes until the mixture starts to foam. Mix the flour, salt and cumin and gradually knead into the yeast mixture with the dough hook on speed 2 until the dough leaves the sides of the bowl.

Continue to knead for 2 minutes until the dough is smooth and elastic. Cover with a damp tea towel and leave to rise in a warm place for 1 hour, or until doubled in volume.

Finely chop the coriander. Knockback the dough, then sprinkle over the coriander. Knead for 2 minutes with the dough hook on speed 2 until the coriander has been incorporated. Divide the dough into 12 portions and roll out each portion to a 15 cm circle. Place on aluminium foil, cover and leave to rise for 1 hour.

Preheat the oven to 260°C/gas mark 9 and place two baking sheets inside to heat up. Transfer the dough circles to the hot baking sheets, sprinkle with water and bake for 8 to 10 minutes. Eat on the day of baking.

# CHERRY TOMATO AND BASIL FOCACCIA

**Makes 2 loaves**

**Prep: 15 minutes**

**Rise: 2 hours**

**Cook: 25 minutes**

**30 g fresh yeast**

**400 ml lukewarm water**

**a pinch of sugar**

**800 g white bread flour**

**1 tbsp salt**

**125 ml olive oil**

**12 sun-dried tomatoes**

**1 bunch of basil**

**24 red cherry tomatoes**

**(or 12 red and 12 yellow cherry tomatoes)**

## extra virgin olive oil and sea salt flakes, to garnish

Crumble the yeast into a measuring jug and add the water. Stir in the sugar and leave for 10 minutes until the mixture starts to foam. Stir in the olive oil. Mix the flour and salt in the mixer bowl with the flat beater on speed 2 for 30 seconds. Change to the dough hook and gradually knead in the yeast mixture until you obtain a soft dough. Knead for 2 minutes on speed 1. Cover with a damp tea towel and leave to rise in a warm place for 1 hour 30 minutes, or until doubled in volume.

Finely chop the sun-dried tomatoes and basil. Knock back the dough and knead with the dough hook on speed 1, mixing in the sun-dried tomatoes and basil. Divide the dough into two portions and roll out each portion to a 1.5 cm thick circle. Place on greased baking sheets and make 12 indentations in each dough circle with your fingertips. Push a cherry tomato into each indentation, cover with a damp tea towel and leave to rise for 30 minutes, or until doubled in volume.

Preheat the oven to 200°C/gas mark 6. Drizzle the dough with olive oil and sprinkle with salt flakes. Bake for 20 to 25 minutes until the focaccias are golden brown and sound hollow when tapped on the base.

## PRETZELS

**Makes 10-12**

**Prep: 20 minutes**

**Rise: 30 minutes**

**Cook: 20 minutes**

**500 g flour**

**a pinch of salt**

**28 g dried yeast**

**40 g sugar**

**75 g softened lard**

**125 ml lukewarm milk**

**125 ml lukewarm water**

**1 egg**

**1 tbsp sesame seeds**

Mix the flour, salt, yeast, sugar and lard in the mixer bowl with the flat beater on speed 2. Knead in the milk and water with the dough hook on speed 2 until you obtain a smooth dough. Cover with a damp tea towel and leave to rise in a warm place for 30 minutes.

Preheat the oven to 200°C/gas mark 6. Knead the dough with the dough hook on speed 1 until it is firm, supple and elastic and leaves the sides of the bowl clean. Divide into 10 or 12 portions. Shape each portion first into a round roll and then into an oval. Roll each oval backwards and forwards, moving your fingers along the dough, to form a strip about 40 cm long, 2.5 cm thick in the middle and 5 mm thick at each end. Pick up the two ends of each strip and make a loop. Cross the ends over twice, then press down on either side of the thickened middle. Repeat with each strip of dough.

Place the pretzels on greased baking sheets. Mix the egg with 1 tablespoon of water and glaze the pretzels. Sprinkle the sesame seeds on top and bake for 20 minutes until golden brown.

# FLACHSWICKEL

**Makes 8**

**Prep: 15 minutes**

**Rise: 1 hour 20 minutes**

**Cook: 30 minutes**

**150 g butter**

**200 ml milk**

**500 g flour**

**7 g dried yeast**

**2 tbsp sugar**

**1 tsp ground cinnamon**

**1 tsp ground cardamom**

**2 eggs**

**a pinch of salt**

**icing sugar, to dust**

Gently melt the butter with the milk in a small saucepan. Pour into the mixer bowl and add the remaining ingredients, except the icing sugar. Knead with the dough hook on speed 2 for 3 minutes until well-blended. Cover and leave to rise in a warm place for 1 hour.

Preheat the oven to 170°C/gas mark 3. Divide the dough into 24 pieces. Roll each piece into a long thin roll about 20 cm long and coat in icing sugar. Lay three rolls side by side and press the top ends together. Plait the three rolls, pressing the bottom ends together to finish. Repeat with the remaining dough rolls. Place on baking sheets and leave for 20 minutes before baking. Bake the flachswickel for 25 to 30 minutes until golden brown and serve warm.

## VANILLA AND POPPY SEED CHALLAH

**Makes 2 loaves**

**Prep: 20 minutes**

**Rise: 2 hours**

**Cook: 35 minutes**

**1 vanilla pod**

**225 ml milk**

**15 g fresh yeast**

**625-700 g flour**

**1 tsp salt**

**75 g melted butter**

**3 tbsp honey**

**3 eggs**

**1 egg yolk**

**2 tsp poppy seeds**

Halve the vanilla pod lengthways and scrape out the seeds. Place these in a pan with the milk and bring to the boil. Remove from the heat and leave until lukewarm. Crumble in the yeast and stir until the yeast has dissolved.

Mix 625 g flour and the salt in the mixer bowl with the flat beater on speed 2 for 15 seconds. Place the dough hook and gradually add the yeast mixture on speed 2. Knead for 1 minute. Add the melted butter, honey and eggs, and knead for another minute. Knead in the remaining 75 g flour, 25 g at a time, until the dough leaves the sides of the bowl clean. Continue to knead on speed 2 for 2 minutes until the dough is smooth and elastic. Place in a greased bowl and coat the dough in the oil. Cover with a damp tea towel and leave to rise in a warm place for 1 hour, or until doubled in volume.

Knock back the dough, then divide into 6 portions. Roll each portion into a 35 cm long thin roll. Plait three rolls together, tuck the ends under and place on a greased baking sheet. Repeat with the remaining portions of dough. Cover with a damp tea towel and leave to rise for 1 hour, or until doubled in volume.

Preheat the oven to 200°C/gas mark 6. Mix the egg yolk with 1 tablespoon of cold water. Glaze the dough and sprinkle with the poppy seeds. Bake for 30 to 35 minutes.

## STOLLEN

**Makes 2 loaves**

**Prep: 2 hours**

**Rise: 2 hours 30 minutes**

**Cook: 30 minutes**

**100 g sultanas**

**50 g currants**

**50 g chopped candied peel**

**grated zest of 1 lemon**

**1 tsp ground cardamom**

**a pinch of black pepper**

**2 tsp vanilla extract**

**2 tbsp dark rum**

**7 g dried yeast**

**1 tsp sugar**

**500 g flour**

**150 ml lukewarm milk**

**½ tsp salt**

**3 tbsp light brown sugar**

**50 g butter**

**2 beaten eggs**

**300 g  marzipan melted butter, to glaze**

**icing sugar, to dust**

Place the sultanas, currants and candied peel in a bowl with the lemon zest, spices and rum. Leave to soak for 2 hours. Mix the yeast, sugar, 125 g flour and the milk in the mixer bowl with the flat beater on speed 2.

Cover with a damp tea towel and leave for 1-hour until frothy.

Knead in the remaining flour, salt, brown sugar, diced butter and most of the beaten eggs with the dough hook on speed 2 until you obtain a manageable dough. Add a little milk, if necessary. Continue to knead for 3 minutes, then cover again and leave to rise for 1 hour, or until doubled in volume.

Knead the dough lightly, then mix in the dried fruit with the dough hook on speed 1; do not overknead. Turn out onto a lightly floured surface and divide into two portions. Roll out each portion into an oval about 1 cm thick. Roll the marzipan into two even-sized rolls to fit the length of the dough. Make an indentation in each portion of dough with a rolling pin and place the marzipan in these indentations. Fold the dough over and seal the edges with some milk. Place the breads on a greased baking sheet, cover with a damp tea towel and leave to rise for 30 minutes.

Preheat the oven to 180°C/gas mark 4. Glaze the dough with milk and bake for 30 minutes until golden brown. Cool on a wire rack, then brush the stollen with melted butter and dust with plenty of icing sugar.

# SAFFRON AND CHOCOLATE BRIOCHE

**Makes 12**

**Prep: 25 minutes**

**Rise: 2 hours 30 minutes**

**Cook: 15 minutes**

**15 g fresh yeast**

**2 tbsp sugar**

**a pinch of saffron threads**

**3 tbsp lukewarm water**

**2 eggs**

**200 g flour**

**a pinch of salt**

**100 g butter**

**100 g chocolate chips**

**1 egg yolk, to glaze**

Crumble the yeast into a measuring jug and add half the sugar, the saffron and water. Stir until dissolved, then leave for 5 to 10 minutes until the mixture starts to foam. Beat in the eggs. Mix the flour, salt and remaining sugar in the mixer bowl with the dough hook on speed 2 for 15 seconds.

Increase to speed 3 and gradually knead in the yeast mixture until you obtain a smooth dough. Continue to knead for 2 to 3 minutes. Add the butter and chocolate chips, then knead for 3 more minutes until well-blended. Place the dough in a greased bowl, cover with a damp tea towel and leave to rise in a warm place for 1 hour 30 minutes, or until doubled in volume.

Grease 12 x 8 cm x 5 cm brioche tins. Knead the dough briefly on a lightly floured surface and divide into 12 portions. Remove a piece of dough about one third the weight of each portion and set aside. Roll the larger portions into balls. Place in the greased tins and make a shallow depression in the top. Roll the smaller portions of dough into balls and place on top of the dough in the tins. Cover and leave to rise for 1 hour.

Preheat the oven to 220°C/gas mark 7. Mix the egg yolk with 1 tablespoon of water. Glaze the brioches and bake for 12 to 15 minutes until golden brown. After baking, immediately turn out the brioches and serve warm with butter.

# MARMALADE DOUGHNUTS

**Makes 18-20**

**Prep: 10 minutes**

**Chill: overnight**

**Rise: 2 hours**

**Cook: 10 minutes**

**250 g flour**

**1 tsp salt**

**25 g sugar**

**15 g fresh yeast**

**100 ml milk**

**1 large egg**

**40 g softened butter**

**100 g fine sugar**

**1 tbsp English Breakfast tea leaves**

**400 g fine-cut orange marmalade**

**sunflower oil, for deep-frying**

Sieve the flour, salt and sugar into the mixer bowl and mix with the dough hook on speed 2 for 10 seconds.

Make a well in the centre. Crumble the yeast into the milk, then pour into the well with the egg. Knead on speed 2 to combine, then increase to speed 4 and continue to knead until the dough forms a ball. Gradually knead in the butter until the dough is smooth. If the dough seems too wet, add a little flour. Place the dough in a greased bowl, cover with clingfilm and chill overnight.

The next day, knock back the dough and knead briefly on a lightly floured surface. Divide the dough into two portions and roll each one out into a long roll 3 to 4 cm thick. Cut each roll into 20 g pieces and place these well apart on greased baking sheets. Cover loosely with oiled clingfilm and leave to rise in a warm place for 1 to 2 hours, or until doubled in volume.

Mix the fine sugar and tea leaves in a coffee mill, then pour onto a large flat plate. Fill a piping bag fitted with a thin nozzle with the marmalade. Heat the sunflower oil in a deep fryer to 190°C. Fry the doughnuts in batches until golden brown. Leave to drain on kitchen paper, then pipe a little marmalade into the doughnuts. Quickly coat the doughnuts in the tea-flavoured sugar and keep warm while you continue to cook and fill the remaining doughnuts. To turn these doughnuts into a dessert, serve them with hot chocolate sauce and a scoop of vanilla ice cream.

# CROISSANTS

**Makes 12-15**

**Prep: 10 minutes**

**Rise: 2 hours**

**Chill: 1 hour 10 minutes**

**Cook: 15 minutes**

**350 g flour**

**¾ tsp salt**

**1 tbsp sugar**

**7 g dried yeast**

**175-200 ml lukewarm milk**

**175 g chilled butter**

**1 beaten egg, to glaze**

Sieve the flour and salt into the mixer bowl. Mix in the sugar and yeast with the dough hook on speed 1 for 15 seconds. Make a well in the flour and pour in the

179

milk. Knead into the flour on speed 1, increase to speed 2 and knead for 2 more minutes until the dough is smooth and elastic. Cover with a damp tea towel and leave to rise in a warm place for 1 hour, or until doubled in volume.

Knock back the dough and knead on a lightly floured surface until smooth. Wrap in a tea towel and chill for 10 minutes. Put the butter between two sheets of clingfilm and roll out into a rectangle. Fold the butter in half and roll out again. Repeat until the butter is pliable but still cold. Flatten to form a 15 cm x 10 cm rectangle.

Turn out the dough onto a floured surface and roll out into a 30 cm x 15 cm rectangle. With a short side facing you, place the butter in the centre of the dough. Fold the bottom third of the dough up over the butter and the top third down over the dough. Press the open sides together to seal. Half turn the dough clockwise.

Repeat the rolling out into a rectangle, folding and turning process twice more. Wrap and chill for 30 minutes. Repeat the rolling out, folding and turning three times, then chill again for 30 minutes.

Dampen two baking sheets with water. Roll out the chilled dough into a 45 cm x 30 cm rectangle. Cut into 6 x 15 cm squares, then slice each square in half diagonally. Starting at the base of each triangle, loosely roll up the dough and fasten the point with a little beaten egg. Arrange the pastries point-side down on the baking sheets, curving the ends to form the characteristic crescent shapes. Lightly brush with beaten egg, cover with oiled clingfilm and leave to rise in a warm place for 1 hour, or until doubled in volume.

Preheat the oven to 220°C/gas mark 7. Brush the croissants again with beaten egg and bake for 3 minutes. Lower the oven temperature to 190°C/gas mark 5 and bake for 10 to 12 minutes until the croissants are golden and crispy. Leave for a few minutes before placing on a wire rack to cool. Serve warm or cold.

# 7.

# DESSERTS

# LIME AND COCONUT ICE CREAM

**Serves 4-6**

**Prep: 15 minutes**

**Freeze: overnight**

**200 ml double cream**

**400 ml coconut milk**

**150 g sugar**

**75 ml lime juice**

**grated zest of 2 limes**

**2 tbsp white rum**

The day before, place the freeze bowl in the freezer. The next day, put the cream, coconut milk and 100 g sugar in a saucepan. Heat gently until the sugar has dissolved, then cool and freeze. Put the remaining sugar, the lime juice and zest in a saucepan. Heat gently until the sugar has dissolved, then reduce until the syrup thickens. Cool and freeze.

When both mixtures are half frozen, mix them together in the mixer bowl with the wire whisk on speed 6 and pour into the freeze bowl. Add the rum and churn with the dasher on speed 2 until almost firm. Spoon into a freezerproof container and place in the freezer overnight to firm up.

# MANGO SORBET WITH CHILLI SYRUP

**Serves 4**

**Prep: 10 minutes**

**Chill: 1 hour Freeze: overnight**

**3 ripe mangoes**

**2 lemons**

**250 g icing sugar**

**Chilli syrup:**

**250 g sugar**

**300 ml water**

**3 large red chillies**

The day before, place the freeze bowl in the freezer. The next day, peel and stone the mangoes, then cut the flesh into chunks. Purée the mango flesh into the mixer bowl with the fruit and vegetable strainer on speed 4. Juice the lemons into the mixer bowl with the citrus juicer on speed 6. Add the icing sugar, then mix with the wire whisk on speed 4 for 2 minutes until the sugar has dissolved and the mixture is smooth. Chill for 1 hour.

Place the mango mixture in the freeze bowl. Attach the dasher and churn on speed 2 until the sorbet is almost firm. Spoon into a freezerproof container and place in the freezer overnight to firm up.

Make the chilli syrup. Bring the sugar and water slowly to the boil, stirring until the sugar has dissolved. Boil for 1 minute. Deseed the chillies and slice the flesh into very thin strips. Add to the syrup and simmer for 25 minutes. Remove from the heat when the chilli strips are translucent and leave to soak overnight.

The next day, scoop the mango sorbet into bowls and drizzle over the chilli syrup.

# TROPICAL FRUIT TEMPURA WITH MATCHA DIPPING SAUCE

**Serves 4**

**Prep: 30 minutes**

**Cook: 10 minutes**

100 g pineapple

1 mango

1 papaya

2 kiwi fruit

2 persimmons (kaki or sharon fruit)

8 physalis (Cape gooseberry)

8 strawberries flour, for dusting icing sugar, to serve

**Dipping sauce:**

125 g sugar

250 ml water

1 tbsp matcha (Japanese powdered green tea)

1 tbsp mirin (sweet Japanese rice wine)

**Tempura batter:**

1 large egg

200 ml iced water

**125 g flour**

**a pinch of salt**

**2 ice cubes**

**groundnut oil, for deep frying**

Peel all the fruit, except the physalis and strawberries. Cut the pineapple, papaya and mango into wedges. Cut the kiwi fruit and persimmon into ½ cm slices. Peel back the leaves of the physalis.

Make the dipping sauce. Bring the sugar slowly to the boil with the water and matcha. Wait until the sugar has dissolved, then simmer the syrup gently until thickened. Stir in the mirin.

Make the batter. Break the egg into the mixer bowl and mix in the water with the wire whisk on speed 4. Change to the flat beater, add the flour and salt and stir into the eggy water on speed 1 until just mixed. Drop the ice cubes into the batter.

Heat the oil to 180°C in a wok or deep fryer. Dust the fruit with a little flour, then dip into the batter. Shake off the excess, then deep fry the fruit in batches in the hot oil for 1 to 2 minutes. Drain on kitchen paper, dust with icing sugar and serve with the dipping sauce.

## ORANGE FLOWER CHURROS WITH HOT CHOCOLATE GANACHE

**Serves 4**

**Prep: 10 minutes**

**Rest: 1 hour**

**Cook: 10 minutes**

**350 g self-raising flour**

**½ tsp ground cinnamon**

**½ tsp salt**

**1 egg**

**300-350 ml milk**

**2 tbsp orange flower water**

**100 g sugar**

**½ tsp Espelette pepper**

**vegetable oil, for deep frying**

**1 quantity Chocolate ganache**

**100 ml hot milk**

**¼ tsp ground allspice**

**grated zest of 1 clementine**

**1-2 tbsp dark rum**

Sieve the flour, cinnamon and salt into the mixer bowl. Make a well in the centre. Whisk the egg with 250 ml milk. Pour into the well and mix with the wire whisk on speed 4. Gradually add enough of the remaining milk on speed 6 until you obtain a smooth batter that can be piped easily. Transfer the batter to a piping bag with a 1 cm star-shaped nozzle.

Heat the oil in a wok or deep fryer to 190°C. Pipe long coils into the hot oil and cook for 4 to 6 minutes until golden brown and cooked. Remove the churros with tongs or a slotted spoon and drain on kitchen paper. Immediately sprinkle with orange flower water. Snip the churros into 10 cm lengths and keep warm while you cook the rest. Mix the sugar and Espelette pepper. When all the churros are cooked, sprinkle them with the spiced sugar.

Make the chocolate ganache as described on page 27 and add the hot milk, allspice and clementine zest. Stir in rum to taste and serve with the churros for dipping.

# SAFFRON MASCARPONE SORBET WITH SYRUPED STAR FRUIT

**Serves 4**

**Prep: 5 minutes**

**Cook: 15 minutes**

**Freeze/marinate: overnight**

**250 g mascarpone**

**1 capsule powdered saffron**

**250 g sugar**

**500 ml milk**

**Syruped star fruit:**

**2 large juicy star fruit (carambola)**

**100 g sugar**

**150 ml water**

**a pinch of saffron threads**

**juice of 1 lemon**

Make this dessert a day beforehand, starting with the mascarpone sorbet. Place all the ingredients in the mixer bowl and beat with the wire whisk on speed 4 until smooth. Pour into the freeze bowl of the ice cream maker and churn with the dasher on speed 2 until almost firm. Spoon into a freezerproof container and freeze overnight until firm.

Slice the star fruit and place in a heavy-based pan with half the sugar. Add the water and bring to the boil, stirring until the sugar has dissolved. Simmer for 5 minutes until the fruit is tender but still holds its shape. Strain and reserve the syrup.

Add the remaining sugar and the saffron to the syrup. Reduce until it becomes sticky. Return the star fruit to the syrup and add the lemon juice. Simmer for 1 minute, then remove from the heat and chill overnight. The next day, bring back to room temperature before serving with the mascarpone sorbet.

# BANANA, MAPLE SYRUP AND RUM ICE CREAM WITH SPICED PINEAPPLE CARPACCIO

**Serves 6**

**Prep: 10 minutes**

**Cook: 10 minutes**

**Freeze/marinate: overnight**

450 g peeled ripe bananas

100 g icing sugar

½ tsp freshly grated nutmeg

3 tbsp maple syrup

3 tbsp dark rum

300 ml double cream

Pineapple carpaccio:

300 g sugar

2 star anise

1 vanilla pod

1 cinnamon stick

½ tsp Szechuan peppercorns

**a pinch of saffron threads**

**400 ml water**

**50 ml white rum**

**1 ripe pineapple**

Make this dessert a day beforehand, starting with the banana ice cream. Purée the bananas into the mixer bowl with the fruit and vegetable strainer on speed 4. Mix in the icing sugar, nutmeg, maple syrup and rum with the flat beater on speed 4 until smooth. Transfer to another bowl.

Clean and dry the mixer bowl thoroughly, then whip the cream with the wire whisk on speed 6 until soft peaks form. Carefully fold into the

banana purée and spoon into a freezerproof container. Freeze overnight.

Make the pineapple carpaccio. Put the sugar in a pan with the star anise, split vanilla pod, cinnamon, Szechuan peppercorns and saffron. Add the water and bring slowly to the boil, stirring until the sugar has dissolved. Simmer for 15 minutes, then stir in the rum. Remove from the heat.

Peel the pineapple and remove the 'eyes'. Quarter the pineapple lengthways, then slice each quarter with the slicing drum on the rotor vegetable slicer/shredder on speed 4. Place the slices in a shallow container and pour over the syrup. Cover and chill overnight. The next day, bring the pineapple back to room temperature before serving with the banana ice cream.

## PLUM AND HIBISCUS FOOL WITH STEM GINGER SHORTBREAD

**Serves 4-6**

**Prep: 5 minutes**

**Infuse: 15 minutes**

**Cook: 35 minutes**

**400 g red plums**

**100-125 g acacia honey**

**1 cinnamon stick**

**300 ml double cream**

**3 tbsp Cointreau**

**2 tbsp icing sugar**

**100 g mascarpone**

**grated zest of 2 clementines**

**1 quantity Stem ginger shortbread, to serve**

**Hibiscus syrup:**

**5 g dried hibiscus flowers**

**250 ml water**

**100 g sugar**

First make the hibiscus syrup. Place the hibiscus flowers in a saucepan with the water. Bring to the boil, immediately remove from the heat, cover and infuse for 15 minutes. Strain, then measure 200 ml hibiscus juice in the pan and add the sugar. Bring slowly to the boil, stirring until the sugar has dissolved. Reduce until syrupy and leave to cool.

Stone the plums and cut them into 2.5 cm chunks. Place in a pan with the honey and cinnamon. Cook for 20 minutes until the fruit is soft and pulpy. Leave to cool.

Whip the cream, Cointreau and icing sugar in the mixer bowl with the wire whisk on speed 6 until soft peaks form. Briefly mix in the mascarpone and clementine zest. Gently fold into the fruit. Spoon into glasses or bowls and drizzle over the hibiscus syrup. Serve with the shortbread.

## SNOW EGGS WITH COFFEE, MAPLE SYRUP AND PECAN CARAMEL

**Serves 4**

**Prep: 30 minutes**

**Cook: 15 minutes**

**4 eggs**

**a pinch of salt**

**250 g sugar**

**500 ml milk**

**1 egg yolk**

**1 tbsp espresso coffee beans**

**2 tbsp maple syrup**

**50 ml cold water**

## 75 g pecan nuts

Separate the eggs and whisk the egg whites with the wire whisk on speed 6 until frothy; reserve the egg yolks. Add the salt and continue to whisk on speed 8 until soft peaks form. Gradually add 50 g sugar, 1 tablespoon at a time, beating well after each addition.

Pour the milk into a shallow saucepan and bring to a gentle simmer. Shape the meringue into quenelles (oval shapes) or make irregularly shaped heaps. Gently lower these into the milk and poach for 2 to 3 minutes, or until they have doubled in size and are firm; do this in batches. Remove the poached meringues with a slotted spoon and leave to drain on kitchen paper. You should have 8 or 12 meringues, depending on their size.

Strain the milk into another pan and bring back to the boil. Place the reserved egg yolks and extra egg yolk into the mixer bowl. Add the finely ground coffee beans and 75 g sugar, then beat with the wire whisk on speed 6 until pale and thick. Pour the milk slowly onto the egg yolks and beat on speed 2 until amalgamated. Pour the mixture back into the pan and heat gently, stirring continuously, until the custard has thickened and coats the back of a wooden spoon. Stir in the maple syrup and leave to cool.

Make the caramel. Put 125 g sugar into a heavy-based saucepan and heat gently until the sugar has dissolved. Raise the heat and boil rapidly until the caramel is a deep golden brown. Swirl the pan from time to time to ensure even browning. Carefully add the water; stand back as the caramel will hiss and spit. Keep stirring on a low heat until the caramel is smooth and slightly runny. Roughly chop the pecans and stir into the caramel.

Spoon the coffee custard into deep plates, adding two or three meringues per plate. Drizzle over the pecan caramel and serve.


# RASPBERRY, ORANGE AND CARDAMOM TIRAMISÙ

**Serves 8**

**Prep: 25 minutes**

**Chill: 4-6 hours**


**8 eggs**

**75 g icing sugar**

**500 g mascarpone**

**grated zest of 1 orange**

**2 tsp ground cardamom**

**3 tbsp orange flower water**

**a pinch of salt**

**500 ml hot espresso coffee**

**100 ml Crème de cacao (chocolate liqueur)**

**30-36 ladyfingers (savoiardi)**

**375 g raspberries**

**4 tbsp cocoa nibs (finely ground cocoa beans)**

Separate the eggs. Mix the egg yolks and icing sugar in the mixer bowl with the wire whisk on speed 8 until pale and thick. Decrease to speed 4 and beat in the mascarpone. Add the orange zest, cardamom and orange flower water. Beat briefly until combined and spoon into a large bowl.

Clean and dry the mixer bowl and wire whisk thoroughly. Whisk the egg whites with the salt on speed 8 until stiff. Carefully fold into the mascarpone mixture.

Stir together the espresso coffee and Crème de cacao. Dip half the ladyfingers into the coffee and use to line the base of a large dish. Cover with half the mascarpone and raspberries.

Dip the remaining ladyfingers into the coffee and layer on top of the raspberries. Cover with the rest of the mascarpone and raspberries. Sprinkle the cocoa nibs over the top. Chill for 4 to 6 hours before serving.

## HAZELNUT AND TONKA BEAN ZABAGLIONE

**Serves 4-6**

**Prep: 10 minutes**

**6 very fresh egg yolks**

**50 g icing sugar**

**1 tsp ground tonka bean**

**6 tbsp Frangelico (hazelnut liqueur)**

**2 tbsp toasted and ground hazelnuts**

Beat the egg yolks, icing sugar, tonka bean and Frangelico in the mixer bowl with the wire whisk on speed 8 for 5 minutes until pale and thick. Increase to speed 10 and whisk for 5 to 7 minutes until the mixture has tripled in volume and is only just pourable. Immediately spoon into 6 serving glasses, sprinkle over the toasted hazelnuts and serve at once.

## DRIED CHERRY, PISTACHIO AND SZECHUAN BISCOTTI

**Makes about 60**

**Prep: 10 minutes**

**Cook: 45 minutes**

**300 g flour**

**1 ½ tsp baking powder**

**150 g sugar**

**1 tsp ground Szechuan pepper**

**3 eggs**

**75 g dried cherries**

**75 g shelled pistachio nuts**

Preheat the oven to 180°C/gas mark 4. Mix the flour, baking powder, sugar and Szechuan pepper in the mixer bowl with the flat beater on speed 2 for 15 seconds.

Change to the dough hook and knead in the eggs on speed 4 until the dough comes together. Add the cherries and pistachio nuts and continue to knead on speed 2 until the dough leaves the sides of the bowl clean. Divide the dough into two portions.

Shape each portion on a lightly floured surface into a 4 to 5 cm wide roll. Place the rolls on a greased baking sheet and bake for 20 to 25 minutes until the edges are golden brown. Remove from the oven and leave for 10 minutes to firm up, then cut the rolls into 1 cm slices and return these to the baking sheet. Bake for 10 minutes, then leave to cool on a wire rack.

# TROPICAL FRUIT AND JASMINE PAVLOVAS

**Serves 8**

**Prep: 10 minutes**

**Cook: 1 hour**

4 egg whites

a pinch of salt

220 g sugar

2 tbsp jasmine tea leaves

2 tsp cornflour

1 tsp white wine vinegar

1 mango

1 papaya

1 star fruit (carambola)

2 kiwi fruit

4 ripe figs

2 x quantity Gin and tonic syllabub, flavoured with 1 vanilla pod and 2 tbsp Midori instead of the gin

1 pomegranate

4 tbsp jasmine syrup

Preheat the oven to 170°C/gas mark 3. Whisk the egg whites and salt with the wire whisk on speed 4 until frothy, then increase to speed 8 until stiff peaks form. Blitz the sugar and jasmine tea in a coffee mill until more or less finely ground. Add to the egg whites, 1 tablespoon at a time. Wait until the previous tablespoon of sugar has been fully incorporated before adding the next one. When the mixture starts to look glossy, increase to speed 10 and continue adding the sugar. When all the sugar has been added, the meringue will be very stiff and glossy.

Remove the wire whisk and sieve the cornflour over the meringue, then sprinkle over the vinegar. Fold both into the meringue with a large metal spoon. Spread the mixture out into 8 x 10 cm circles on two silicone sheets. Make a depression in the meringues, to contain the filling later. Place the meringues in the oven, then immediately reduce the heat to 130°C/gas mark 1/2. Bake the meringues for 1 hour.

The meringues are ready when they feel dry to the touch and can easily be lifted off the silicone sheets. Turn the oven off but leave the meringues in the oven with the oven door ajar until they have cooled completely. This will prevent the meringues from cracking.

Peel the fruit if necessary and slice into wedges or cut into slices. Spoon the syllabub into the meringues and top with the fruits. Halve the pomegranate and scatter the seeds all over the fruit. Drizzle with the jasmine syrup and serve at once.

# WHITE CHOCOLATE AND GREEN TEA BAVAROIS WITH ELDERFLOWER GRANITA

**Serves 4-6**

**Prep: 20 minutes**

**Cook: 30 minutes**

**Freeze: 4 hours**

**100 ml milk**

**300 ml double cream**

**1 tbsp matcha (Japanese powdered green tea)**

**50 g sugar**

**3 large egg yolks**

**100 g white chocolate**

**2 gelatine leaves**

**Elderflower granita:**

**150 g sugar**

**400 ml water**

**125 ml elderflower cordial**

**juice of ½ lemon**

Put the milk, 100 ml cream and matcha in a saucepan with ½ tablespoon of sugar and bring to the boil, stirring until the matcha and sugar have dissolved. Whisk the egg yolks and remaining sugar in the mixer bowl with the wire whisk on speed 4 until pale and thick. Pour the hot milk onto the egg yolks and whisk on speed 2 until well-blended.

Return the mixture to the pan and stir on a gentle heat until the custard thickens and coats the back of a spoon. Grate the white chocolate and soak the gelatine in cold water. When the custard has thickened, remove from the heat and pour onto the chocolate. Stir until smooth, then squeeze out the gelatine and mix into the white chocolate custard. Strain through a sieve into a bowl and cover the surface with clingfilm to prevent a skin from forming on top. Chill until the custard has cooled completely and begun to set.

Make the elderflower granita. Put the sugar in a pan with 150 ml water. Bring slowly to the boil, stirring until the sugar has dissolved. Remove from the heat and pour 250 ml sugar syrup into a measuring jug. Add the elderflower cordial, lemon juice and 250 ml water. Mix well, then pour into a shallow freezer-proof container. Freeze for 2 hours until the granita is firm around the edges.

Meanwhile, whip the remaining 200 ml cream in the mixer bowl with the wire whisk on speed 6 until soft peaks form. Carefully fold into the white chocolate custard and divide over cups or glasses. Chill until set. Or leave to set in a large bowl and scoop out quenelles of bavarois before serving.

When the granita is firm around the edges, break up the ice crystals with a fork and stir them into the granita. Return to the freezer for 30 minutes and fork through again. Repeat this freezing and forking through until the granita is fluffy, then serve with the bavarois.

# PETITS POTS DE CRÈME AU CHOCOLAT ET À LA MENTHE

**Serves 4**

**Prep: 5 minutes**

**Infuse: 20 minutes**

**Cook: 10 minutes**

**Chill: 2 hours**

**300 ml single cream**

**½ bunch of fresh mint**

**200 g dark chocolate (70%)**

**2 large egg yolks**

**3 tbsp Crème de menthe (mint liqueur)**

**25 g butter**

**crystallized mint leaves, to decorate**

**8 biscuits of your choice, to serve (optional)**

Place the cream and mint in a saucepan and bring to the boil. Immediately remove from the heat, cover and infuse for 20 minutes. Strain the cream back into the saucepan, discarding the mint.

Grate the chocolate into the mixer bowl with the coarse shredding drum on the rotor vegetable slicer/shredder on speed 4. Bring the infused cream back to the boil and pour over the chocolate. Stir until smooth, then mix in the egg yolks and Crème de menthe with the wire whisk. Leave to cool slightly, then mix in the butter as well. Pour into small pots or glasses and chill. Bring back to room temperature and decorate with crystallized mint leaves before serving with the biscuits.

## CHOCOLATE MOUSSE WITH LAPSANG SOUCHONG AND SINGLE MALT WHISKY

**Serves 4**

**Prep: 15 minutes Chill: overnight**

**150 g dark chocolate (70%)**

**50 g butter**

**50 g sugar**

**½ tbsp Lapsang Souchong tea leaves**

**3 large eggs**

**2 tbsp single malt whisky**

**a pinch of salt**

**cocoa nibs (coarsely ground cocoa beans), to serve**

Grate the chocolate into a heatproof bowl with the coarse shredding drum on the rotor vegetable slicer/shredder on speed 4. Add the butter and melt over a pan of simmering water. Stir until smooth, then set aside.

Blitz the sugar and tea in a coffee mill until finely ground. Separate the eggs. Beat the egg yolks, tea sugar and whisky in the mixer bowl with the wire whisk on speed 6 until pale and thick. Gently fold in the melted chocolate.

Clean and dry the mixer bowl and wire whisk thoroughly, then whisk the egg whites and salt on speed 8 until stiff. Carefully fold into the chocolate mixture. Spoon into bowls or onto plates and chill overnight; or pour into a large bowl and serve the mousse in the shape of quenelles. Just before serving, sprinkle over the cocoa nibs.

# BLACK FOREST FONDANTS

**Serves 8**

**Prep: 30 minutes**

**Cook: 15 minutes**

**250 g dark chocolate (70%)**

**250 g butter**

**2 tbsp Kirsch**

**225 g sugar**

**2 tonka beans**

**5 eggs**

**5 egg yolks**

**50 g flour**

**Cherry compote:**

**800 g tinned morello cherries**

**4 tbsp Kirsch**

**Mascarpone cream:**

**1 vanilla pod**

**250 g mascarpone cheese**

**50 ml milk**

**200 g soured cream**

**2 tbsp Kirsch**

**50 g icing sugar**

First make the cherry compote. Drain the cherries, reserving the juices, and place half the cherries in a bowl. Pour over the Kirsch and leave to marinate for 20 minutes. Afterwards, drain the cherries and set aside. Add the Kirsch to the reserved cherry juices. Bring these to the boil and reduce to a thick syrup. Mix into the remaining cherries and set aside.

Make the mascarpone cream. Split the vanilla pod and scrape out the seeds. Stir these into the mascarpone with the milk. Mix the soured cream, Kirsch and the icing sugar with the wire whisk on speed 6. Fold into the vanilla mascarpone and chill.

Preheat the oven to 200°C/gas mark 6. Melt the chocolate and butter with the Kirsch. Blitz the sugar and tonka beans to a fine powder in a coffee mill. Mix the eggs, egg yolks and tonka sugar in the mixer bowl with the wire whisk on speed 8 until pale and thick. Carefully fold in the melted chocolate. Sieve the flour over the batter and mix briefly.

Divide half the chocolate batter over 8 x 200 ml well-greased and floured dariole moulds. Gently place 3 Kirsch-marinated cherries on top, then cover with the remaining batter. Bake for 10 to 12 minutes until the tops feel dry but the centre is still soft. Rest for 1 minute before turning out and serving with the cherry compote and mascarpone cream.

## CHOCOLATE, APRICOT AND SZECHUAN PEPPER TARTLETS

**Serves 8**

**Prep: 20 minutes**

**Chill: 1 hour**

**Cook: 30 minutes**


**100 g softened butter**

**100 g icing sugar**

**2 large egg yolks**

**225 g flour**

**25 g cocoa powder**

**a pinch of salt**

**soured cream, to serve cocoa powder, for dusting**

**Chocolate filling:**

**300 g dried apricots**

**75 ml water**

**3 tbsp lemon juice**

**200 g good-quality milk chocolate**

**100 g dark chocolate (85%)**

**200 g butter**

**2 eggs**

**2 egg yolks**

**50 g sugar**

**1 ½ tsp freshly ground Szechuan pepper**

Cream the butter and sugar in the mixer bowl with the flat beater on speed 2 until pale and fluffy. Gradually beat in the egg yolks until well- blended. Mix in the flour, cocoa powder and salt on speed 4 until the pastry leaves the sides of the bowl. Wrap in clingfilm and chill for 30 minutes. Roll out the pastry on a lightly floured surface and line 8 x 8 cm greased tartlet tins. Prick the base with a fork and chill for 30 minutes.

Preheat the oven to 180°C/gas mark 4. Line the tartlet cases with baking paper and fill with baking beans. Bake blind for 20 minutes. Remove the paper and beans and leave to cool.

Meanwhile, make the apricot purée. Finely chop the apricots and simmer for 5 minutes with the water. Mix in the blender on purée speed    with the lemon juice until smooth. Spread the apricot purée in the cooled tartlet cases.

Melt the chocolate and butter together. Mix the eggs, egg yolks, sugar and Szechuan pepper in the mixer bowl with the wire whisk on speed 8 until pale and thick. Gently fold in the melted chocolate and spoon into the tartlet cases. Bake for 8 minutes at 180°C/gas mark 4. The filling will not have set completely but will continue to set as it cools. Serve the tartlets warm or at room temperature with soured cream. Dust with cocoa powder before serving.

# CHOCOLATE AND LAVENDER PITHIVIERS

**Serves 6**

**Prep: 15 minutes**

**Chill: 30 minutes**

**Cook: 25 minutes**

**50 g marzipan**

**75 g icing sugar**

**2 tbsp dried lavender flowers**

**50 g softened butter**

**2 egg yolks**

**2 tbsp cocoa powder**

**50 g ground almonds**

**2 tbsp Crème de cacao (chocolate liqueur)**

**1 quantity Pastry cream, flavoured with 50 g grated dark chocolate (70%)**

**1 quantity Flaky pastry**

**1 egg**

**1 tbsp icing sugar**

**1 quantity Vanilla custard, flavoured with 2 tbsp dried lavender flowers, to serve**

Grate the marzipan into the mixer bowl with the medium shredding drum on the rotor vegetable slicer/shredder on speed 4. Blitz the icing sugar and lavender in a coffee mill. Add to the mixer bowl with the butter. Beat with the flat beater on speed 2 until pale and fluffy. Beat in the egg yolks one at a time, then mix in the cocoa powder, almonds and Crème de cacao. Finally, beat in 100 ml pastry cream 1 tablespoon at a time (keep the remaining pastry cream for another recipe). Cover and chill.

Roll out the pastry on a lightly floured surface. Cut out 12 x 10 cm circles and place six of these on a greased baking sheet lined with greaseproof paper. Spread a sixth of the filling on each pastry circle, leaving a border of 2 cm all around. Brush these borders with water, then place the remaining pastry circles on top of the filling. Press the edges together to seal, then make a fluted pattern around the edges with a fork. Chill for 30 minutes.

Preheat the oven to 220°C/gas mark 7. Beat the egg with 1 tablespoon of water and glaze the pastries. With a knife score faint curved lines from the centre of the pastries to the edges, making a cartwheel pattern. Make a small hole in the top to allow steam to escape, then bake the pithiviers for 20 minutes or until well-risen and golden brown. Dust the tops with icing sugar and return to the oven for another 5 minutes. Remove the pithiviers from the oven and leave to cool for 10 minutes before serving with lavender custard.

## LEMON AND VANILLA TART

**Serves 8**

**Prep: 15 minutes**

**Cook: 1 hour 10 minutes**

**1 quantity Shortbread pastry, made with 2 tbsp dried lavender flowers**

**2-3 lemons**

**1 vanilla pod**

**4 eggs**

**1 egg yolk**

**200 g sugar**

**200 ml double cream**

**icing sugar, to dust (optional)**

**clotted cream (or mascarpone), to serve**

Preheat the oven to 180°C/gas mark 4. Roll out the pastry on a lightly floured surface and use to line a greased 23 cm loose-based tart tin. Prick the pastry with a fork and fill with greaseproof paper and baking beans. Bake blind for 12 minutes. Remove the paper and beans, and bake for another 15 minutes.

Meanwhile, grate the zest from the lemons and set aside. Cut the lemons in half and squeeze the juice with the citrus juicer on speed 6. Split the vanilla pod and scrape out the seeds. Whisk the eggs, egg yolk, sugar, lemon zest and vanilla seeds in the mixer bowl with the wire whisk on speed 6 until pale and thick. Stir in the lemon juice on speed 2, then gently fold in the cream. Pour into the hot tart case and reduce the oven temperature to 130°C/gas mark 1/2. Bake the tart for 40 minutes until the filling feels just firm to the touch. Leave to cool before dusting with icing sugar, if you like. Serve with clotted cream or mascarpone.

# HAZELNUT TARTLETS WITH RED WINE PEARS

**Serves 6-8**

**Prep: 10 minutes**

**Chill: 30 minutes**

**Cook: 1 hour 15 minutes**

**1 quantity Sweet shortcrust pastry**

**125 g softened butter**

**125 g sugar**

**2 beaten eggs**

**25 g flour**

**125 g ground hazelnuts**

**1 quantity White chocolate ice cream, to serve**

**Red wine pears:**

**½ bottle Beaujolais**

**125 ml red port**

**250 ml freshly squeezed orange juice**

**zest of 2 oranges**

**zest of 1 lemon**

**1 bay leaf**

**2 tsp coriander seeds**

**1 sprig of rosemary**

**1 cinnamon stick**

**1 tonka bean**

**250 g sugar**

**3-4 ripe but firm stewing pears**

First make the red wine pears. Place all the ingredients, except the pears, in a large saucepan and slowly bring to the boil. Stir until the sugar has dissolved. Reduce the heat to a simmer and cook for 10 minutes. Peel and core the pears but leave the stalks on. Halve the pears lengthways and lower into the syrup. Cover with greaseproof paper and gently poach for 20 to 30 minutes until the pears are tender. Remove from the syrup and leave to cool. Reduce the poaching liquor until syrupy, then strain and cool.

Preheat the oven to 180°C/gas mark 4. Grease 6 to 8 x 10 cm tartlet tins. Roll out the pastry and line the tartlet tins, then chill for 30 minutes.

Cream the butter and sugar in the mixer bowl with the flat beater on speed 4 until pale and fluffy. Gradually beat in the eggs on speed 6 until well-blended. Fold the flour into the creamed mixture on speed 2, then fold in the hazelnuts. Spoon the frangipane into the tartlet tins.

Slice the pears lengthways into thin slices, so you can fan them out by gently pushing down on the pears. Push a pear half into each tartlet and bake for 30 to 35 minutes until the frangipane has set. Leave to cool on a wire rack and serve lukewarm or at room temperature with the ice cream and a drizzle of red wine syrup. You could also serve the tartlets with the white chocolate custard before churning it to ice cream.

## APPLE AND THYME PIE WITH CHEDDAR CRUST

**Serves 8**

**Prep: 15 minutes**

**Chill: 50 minutes**

**Cook: 30 minutes**

350 g flour

a pinch of salt

1 tbsp icing sugar

60 g butter

60 g very cold lard

60 g mature Cheddar

2 beaten eggs

2-3 tbsp milk

1 quantity Vanilla custard, flavoured with ½ tsp freshly grated nutmeg instead of vanilla, to serve

**Apple filling:**

750 g eating apples, eg Cox's Orange Pippin, Gala or Pink Lady

1.25 kg cooking apples

50 g butter

100 g sugar

grated zest of 2 oranges

grated zest of 2 lemons

2 tbsp finely chopped thyme

1 heaped tbsp semolina

Apple pie is perhaps the most popular fruit pie and it comes in many guises. In this recipe, I have introduced a savoury note with the Cheddar cheese. A surprising addition which works remarkably well with the apples and thyme.

Sieve the flour, salt and icing sugar into the mixer bowl. Dice the butter and lard, and grate the Cheddar with the medium shredding drum on the rotor vegetable slicer/shredder on speed 4. Mix the butter, lard, Cheddar and 1 egg into the flour with the flat beater on speed 2. Add the milk and continue to mix until the dough comes together and leaves the sides of the bowl. Knead briefly on a lightly floured surface, wrap in clingfilm and chill for 30 minutes.

Peel, core and slice the apples with the slicing drum on the rotor vegetable slicer/shredder on speed 4. Melt the butter in a large saucepan and cook the apples with the sugar for 5 minutes until they begin to soften. Remove from the heat and strain the apples, reserving the juices. Stir the orange and lemon zest and the thyme into the apples.

Divide the pastry into two portions, one twice as large as the other. Roll out the larger portion on a lightly floured surface and use to line a deep 18-20 cm x 8 cm springform tin. Prick the pastry with a fork and chill for 20 minutes.

Preheat the oven to 200°C/gas mark 6. Line the prepared tin with greaseproof paper and baking beans. Bake blind for 25 minutes. Remove the paper and beans, and bake for another 5 to 10 minutes.

Increase the oven temperature to 220°C/gas mark 7. Sprinkle the semolina over the pastry and pile in the apples. Roll out the remaining pastry to a circle slightly larger than 18 or 20 cm. Place on top of the apples, trim the edges and press lightly together. Glaze the top with the remaining egg and cut a small hole in the centre for the steam to escape during cooking. Bake for 25 to 30 minutes until golden brown. Leave to rest for 20 minutes before serving with the strained apple juices and nutmeg custard.

# ABOUT THE AUTHOR

**Barbara Colbert** is passionate writer about empowering people to lead active healthy lifestyles by teaching them the personalized skills they need to fuel themselves with whole foods while maintaining a healthy life balance.

Printed in Great Britain
by Amazon

81258377R00129